THE MANAGEMENT AND HEALTH OF FARMED DEER

Current Topics in Veterinary Medicine and Animal Science

Recent publications

1984

26. Manipulation of Growth in Farm Animals, edited by J.F. Roche and D. O'Callaghan. ISBN 0-89838-617-8
27. Latent Herpes Virus Infections in Veterinary Medicine, edited by G. Wittmann, R.M. Gaskell and H.-J. Rziha. ISBN 0-89838-622-5
28. Grassland Beef Production, edited by W. Holmes. ISBN 0-89838-650-0
29. Recent Advances in Virus Diagnosis, edited by M.S. McNulty and J.B. McFerran. ISBN 0-89838-674-8
30. The Male in Farm Animal Reproduction, edited by M. Courot. ISBN 0-89838-682-9

1985

31. Endocrine Causes of Seasonal and Lactational Anestrus in Farm Animals, edited by F. Ellendorff and F. Elsaesser. ISBN 0-89838-738-8
32. Brucella Melitensis, edited by J.M. Verger and M. Plommet. ISBN 0-89838-742-6

1986

33. Diagnosis of Mycotoxicoses, edited by J.L. Richard and J.R. Thurston. ISBN 0-89838-751-5
34. Embryonic Mortality in Farm Animals, edited by J.M. Sreenan and M.G. Diskin. ISBN 0-89838-772-8
35. Social Space for Domestic Animals, edited by R. Zayan. ISBN 0-89838-773-6
36. The Present State of Leptospirosis Diagnosis and Control, edited by W.A. Ellis and T.W.A. Little. ISBN 0-89838-777-9
37. Acute Virus Infections of Poultry, edited by J.B. McFerran and M.S. McNulty. ISBN 0-89838-809-0

1987

38. Evaluation and Control of Meat Quality in Pigs, edited by P.V. Tarrant, G. Eikelenboom and G. Monin. ISBN 0-89838-854-6
39. Follicular Growth and Ovulation Rate in Farm Animals, edited by J.F. Roche and D. O'Callaghan. ISBN 0-89838-855-4
40. Cattle Housing Systems, Lameness and Behaviour, edited by H.K. Wierenga and D.J. Peterse. ISBN 0-89838-862-7
41. Physiological and Pharmacological Aspects of the Reticulo-rumen, edited by L.A.A. Ooms, A.D. Degryse and A.S.J.P.A.M. van Miert. ISBN 0-89838-878-3
42. Biology of Stress in Farm Animals: An Integrative Approach, edited by P.R. Wiepkema and P.W.M. van Adrichem. ISBN 0-89838-895-3
43. Helminth Zoonoses, edited by S. Geerts, V. Kumar and J. Brandt. ISBN 0-89838-896-1
44. Energy Metabolism in Farm Animals: Effects of Housing, Stress and Disease, edited by M.W.A. Verstegen and A.M. Henken. ISBN 0-89838-974-7
45. Summer Mastitis, edited by G. Thomas, H.J. Over, U. Vecht and P. Nansen. ISBN 0-89838-982-8

1988

46. Modelling of Livestock Production Systems, edited by S. Korver and J.A.M. van Arendonk. ISBN 0-89838-373-0
47. Increasing Small Ruminant Productivity in Semi-arid Areas, edited by E.F. Thomson and F.S. Thomson. ISBN 0-89838-386-2
48. The Management and Health of Farmed Deer, edited by H.W. Reid. ISBN 0-89838-408-7

The Management and Health of Farmed Deer

A Seminar in the CEC Programme of Coordination of Research
in Animal Husbandry, held in Edinburgh on 10–11 December 1987

Sponsored by the Commission of the European Communities,
Directorate-General for Agriculture,
Division for the Coordination of Agricultural Research

Edited by

H.W. REID

Moredun Research Institute,
Edinburgh, Scotland,
United Kingdom

KLUWER ACADEMIC PUBLISHERS
DORDRECHT / BOSTON / LONDON
FOR THE COMMISSION OF THE EUROPEAN COMMUNITIES

Library of Congress Cataloging in Publication Data

The Management and health of farmed deer / edited by H.W. Reid.
      p.   cm. -- (Current topics in veterinary medicine ; 48)
    "A seminar in the CEC Programme of Coordination of Research in
  Animal Husbandry, held in Edinburgh on 10-11 December 1987;
  sponsored by the Commission of European Communities, Directorate
  -General for Agriculture, Division for the Coordination of
  Agricultural Research."
    Includes index.

    1. Deer farming--Congresses. 2. Deer--Diseases--Congresses.
  I. Reid, H. W. II. Commission of the European Communities.
  III. CEC Programme of Coordination of Research in Animal Husbandry.
  IV. Commission of the European Communities. Division for the
  Coordination of Agricultural Research. V. Series: Current topics in
  veterinary medicine and animal science ; v. 48.
  SF401.D3M36 1988
  636.2'94--dc19   88-13474
                                                             CIP

ISBN-13: 978-94-010-7090-4   e-ISBN-13: 978-94-009-1325-7
DOI: 10.1007/978-94-009-1325-7

Publication arrangements by
Commission of the European Communities
Directorate-General Telecommunications, Information Industries and Innovation, Scientific and
Technical Communications Service, Luxembourg

EUR 11540

© 1988 ECSC, EEC, EAEC, Brussels and Luxembourg

Softcover reprint of the hardcover 1st edition 1988

LEGAL NOTICE
Neither the Commission of the European Communities nor any person acting on behalf of the
Commission is responsible for the use which might be made of the following information.

Published by Kluwer Academic Publishers,
P.O. Box 17, 3300 AA Dordrecht, The Netherlands.

Kluwer Academic Publishers incorporates the publishing programmes of
D. Reidel, Martinus Nijhoff, Dr W. Junk and MTP Press.

Sold and distributed in the U.S.A. and Canada
by Kluwer Academic Publishers,
101 Philip Drive, Norwell, MA 02061, U.S.A.

In all other countries, sold and distributed
by Kluwer Academic Publishers Group,
P.O. Box 322, 3300 AH Dordrecht, The Netherlands.

CONTENTS

SESSION V

Chairman: J.A. Milne

PREFACE

The farming of deer as an alternative to traditional livestock enterprises is now firmly established and is expanding within several countries of the European Economic Community. However, the successful farming of deer requires the adoption of appropriate management schemes to accommodate the biological requirements of these animals. Much experience has now been gained and it is essential that this information becomes readily available througout the Community. In addition, as the volume of deer farming has increased a number of health problems have become recognised which present features distinct from other domestic ruminants. Although knowledge is still incomplete it would appear that deer may react to certain pathogens in a very different way to other domestic ruminants, presenting new problems of diagnosis and control. The rapid detection of these conditions and development of appropriate control strategies will be essential for the establishment of an economically viable deer farming industry in the Community.

Much of the information on the management of farmed deer and their diseases is anecdotal and fragmented and the purpose of this meeting was to accelerate the dissemination of this knowledge between scientists in the Community committed to the development of this area of agricultural industry.

The meeting, financed by the Commission of the European Communities from its budget for the Coordination of Agricultural Research in the Community was held in Scotland, on the 10th to 11th December, 1987. The organisation was shared by the Moredun Research Institute and the Macaulay Land Use Research Institute both recognised in the Community as leading centres for research into the health and management of farmed deer. The work of both Institutes supported by the Department of Agriculture and Fisheries for Scotland, has been instrumental in bringing deer farming in Scotland to its present advanced state and in ensuring the future expansion of the industry in Europe. The Commission of the European Communities would like to record thanks to the Chairmen of sessions for ensuring that the programme was adhered to and the staff of the Moredun Research Institute and the Macaulay Land Use Research Institute who assisted in the smooth running of the conference.

December 1987

Participants (left to right) at the first Conference sponsored by the CEC on the farming of deer, at the Moredun Research Institute

P.P. Pastoret (Belgium); J. Barrat (France); J. Winkelmann (Germany); G. Reinken (Germany); E. Thiry (Belgium); M. Prave (France); R.H. Seed (UK); J. Fletcher (UK); J.M. Scudamore (UK); F. Vigh Larsen (Denmark); H.W. Reid (UK); T.L. Alexander (UK); M. O'Toole (Ireland); J. Connell (CEC); R.J. Jorgensen (Denmark); H. Hemmer (Germany); J.A. Milne (UK); F.H.M. Borgsteede (Netherlands); M.E. Brown (UK); P.F. Nettleton (UK); H.V. Krogh (Denmark); F. Stuart (UK); C. Ek-Kommonen (Finland); R. Munro (UK); D. Buxton (UK).

MANAGEMENT AND NUTRITION OF FARMED RED DEER

J.A. MILNE

Macaulary Land Use Research Institute,
Bush Estate, Penicuik, Midlothian, EH26 0PY

INTRODUCTION

Although deer have been kept in deer parks in Europe since the Middle Ages, predominantly for hunting, no attempts were made to manage them commercially for meat production. However, in the last 15 years the farming of the red deer has developed both in the UK and in New Zealand and there is now considerable interest being shown in many countries of Europe.

The initial stimulus for the growth of New Zealand deer farming was the rapid development of a market for antler velvet, which is used as a tonic in South East Asia. The farmed deer industry increased rapidly through the capture of large numbers of the wild population, mainly found in the South Island. Because of changes in the price ratios of antler velvet and meat, production from the 240,000 deer farmed in New Zealand (1983 figures) is now orientated to the supply of both commodities.

In the UK the number of farmed red deer has increased more slowly with 13,000 breeding hinds being farmed in 1987. Although initially some animals were captured from the wild, a high proportion of the current population of breeding hinds have been born on deer farms. The purpose of deer farming in the UK is solely the production of meat. Systems of farming have developed in a stratified manner in the UK, akin to the structure of its sheep industry. Weaned calves are produced in the upland areas of the UK and are transferred to lowland areas, where systems of producing animals for slaughter and more intensive systems of breeding and production of slaughter animals have developed.

Research has been conducted on the farming of red deer in the UK and New Zealand since 1970. Most of the initial research in the UK was conducted jointly by the Hill Farming Research Organisation and the Rowett Research Institute in Scotland and the findings have been reported in two reports entitled "Farming the Red Deer" (Blaxter et al, 1974; Blaxter et al, 1988). There is no such record of the research that has been conducted in New Zealand but the Proceedings of an

International Conference on the Biology of Deer Production (1985) describe much of the research conducted in New Zealand. The object of this paper is to summarise the research described in more detail in these publications and to discuss new work which has increased our understanding and which has allowed improved systems of production to be synthesised.

SIZE OF RED DEER

There are substantial differences in the liveweights of wild red deer from various locations in Europe. Stags in the forests of central and eastern parts of Europe may exceed 300 kg in liveweight, whilst those in Scotland at the north-western limit of their natural range weigh only 150 kg. Some of these differences can be ascribed to differences in nutrition and environment rather than inheritance. Evidence for this is provided by the liveweight of stags of the same genotype. Stags farmed on hill areas of Scotland have a mature liveweight of 145 kg. Male calves, the offspring of these stags, when penned indoors and given a high quality diet, reached a live weight of 190 kg at 3 years of age (Blaxter et al, 1974).

Breeding objectives have concentrated on selecting animals of good temprament and on increasing calf growth rates. Adoption of this latter objective is likely to lead to an increase in the size of farmed red deer. There is also considerable interest in the UK and New Zealand in the crossing of red deer hinds with large stags imported from Central Europe and of hybridising with sub species such as the larger Waipiti. This also has the objective of increasing liveweight gains of calves and is also likely to lead to an increase in the mature size of the breeding herd of farmed red deer.

REPRODUCTION

Puberty in the female red deer can be achieved at 16 months of age. The probability of calving is predominantly determined by liveweight rather than by age (Blaxter and Hamilton, 1980). At 50 kg the probability of having a calf is zero; it increases to 50% at a liveweight of 60 kg and 80% at 70 kg. It is thought that young hinds like their older counterparts, are mated and conceive at their first oestrus. Recent evidence confirms these observations (Hamilton, personal communication). Yearling hinds with a low probability of calving, according to their liveweight, when given an increased plane of nutrition during the mating period so that their liveweight increased, showed first oestrus at a later date, reflecting their increased liveweight, and conceived at first oestrus.

The probability of having a calf and the time of calving is also related to the adult hind liveweight. In relation to time of calving, for each 4 kg heavier a hind is at mating, calving will take place one day earlier (Blaxter and Hamilton, 1980).

The red deer is a seasonal breeder with the onset of the breeding season occurring at the end of September in the UK. Oestrous activity continues until the following March, if hinds are not mated. A continued lactation has been shown to delay the onset of the breeding season by 8 days; hinds which were weaned in mid-September prior to the introduction of stags were compared with hinds weaned in November (Milne *et al*, 1987). This delay has been associated with the effect of the suckling stimulus on plasma prolactin concentrations (Loudon *et al*, 1983).

The daily dosing of hinds with 5 mg of melatonin, given in the late afternoon, has been found to bring forward the date of onset of oestrus by 25-35 days (Adam *et al*, 1985) and calves have been born in late April rather than at the normal time in early June. This has advantage in areas where there is a long season of herbage growth allowing a longer period of calf growth from inexpensive nutrients derived from pasture.

Melatonin also alters the coincident physiological cycles of intake and coat growth as well as that of reproduction. The natural decline in intake which occurs in the autumn occurs earlier in melatonin-treated hinds and there appears to be a restricted development of the winter coat (J.A. Milne, A.S.I. Loudon, J. Curlewis, A.S. McNeilly and A.M. Sibbald, unpublished data). These factors may require to be considered before Melatonin is used in some situations.

Stags reared under farm conditions are usually fertile at 16 months of age but are not used until they are over two years of age. Stags will successfully mate 20-30 hinds in single sire groups and also when they are with other stags in an enclosure, provided that there is sufficient space to allow mating groups to remain apart and that the stags are close together in the dominance order (Hamilton and Blaxter, 1980).

Red deer have a low reproductive rate, producing only one calf per annum. Natural twinning is very rare and attempts to induce twinning by exogenous hormonal means have been relatively unsuccessful to date.

The lifetime reproductive performance of farmed red deer has yet to be established but hinds born in 1970-72 have produced high calving rates in the Macaulay Land Use Research

Table 1   The lifetime performance of red deer

	1970-86	1987
Number of hinds	70	50
Mortality rate per annum (%)	1.2	10
Weaning rate of calves (%)	85	73

Institute's herd at the Glensaugh Research Station until 1986, although there is some evidence that performance is now declining (Milne *et al*, 1987), as can be seen in Table 1 above.

PREGNANCY

The pattern of deposition of nutrients in the foetus of red deer hinds has been described by Adam *et al*. (1987) and is similar to that described for other domesticated ruminants. Little information exists on the effect of undernourishment in pregnancy on foetal mortality or birth weight. Experience would suggest that losses of liveweight of 5% over pregnancy have no detrimental effect. However since winter feed costs are a major determinant of total costs in deer farming systems, there is a need to provide more quantitative information on nutrition in pregnancy to improve the precision of winter feeding.

Birth weights of calves range from 6-9 kg depending on the sex of the calf, the liveweight of the hind at mating and the nutrition of the hind, particularly in late pregnancy. Male calves are approximately 0.5 kg heavier than female calves. Every 2 kg increase in liveweight of hinds at mating increases calf birth weight by 0.1 kg and grazing hinds on pasture of higher quality during the winter increased birth weight by 0.5 kg (Blaxter *et al*, 1988). Perinatal calf mortality ranges from 5-10% in farmed red deer (Blaxter *et al*, 1988). It is related to birth weight with calves of less than 4 kg having 100% mortality and those of 7-8 kg having 5%.

LACTATION AND CALF GROWTH

The lactation curve of red deer and the composition of the milk was first described by Arman *et al*. (1974) with housed red deer. They showed that the milk yield was 1.7 kg in the first month of lactation and that the milk was rich in fat (8-13%) and protein (7-9%) and contained about 4.5% lactose. Loudon *et al* (1984) demonstrated that the lactating hind had the ability to respond to improved levels of nutrition as provided by a sown species sward compared to a hill sward.

Over a 100-day lactation period the milk yield of the hinds on the sown sward was 60% greater than that on the hill sward. These differences in lactation were associated with differences in grazing behaviour. Hinds on the hill area had slower biting rates than hinds on the sown ryegrass-clover pasture and they grazed for about twice as long each day throughout the period of lactation. Evidently their lactation and the growth of their calves was limited by the scarcity of good quality herbage on the hill. On sown swards calf growth rates between birth and weaning are also greater at higher sward heights and

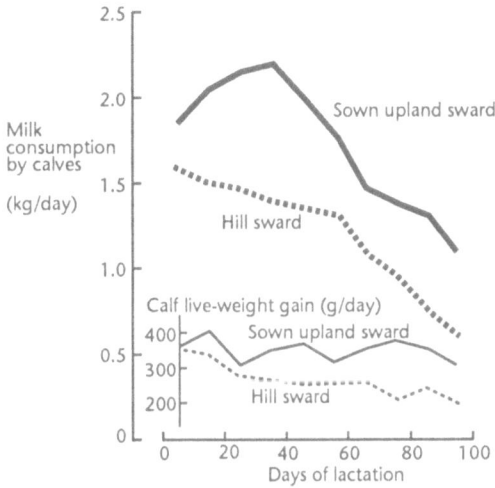

FIGURE 1. Performance of red deer during lactation on contrasting swards

herbage masses (Loudon *et al*, 1984). Depending upon the amount and quality of herbage available calf growth rates in this period range from 250-400 g/day. After weaning, calf growth rates have also been found to be positively related to herbage mass of pasture but liveweight gains are lower (50-250 g/day) than prior to weaning (Milne *et al*, 1987). As autumn proceeds, liveweight gains of grazing calves decline. This appears to be associated with weather-stress rather than nutrient availability and Yerciniosis, a stress-related bacterial disease, has been observed in weaned calves both under UK and New Zealand conditions (Henderson, 1983). Under the sward and climatic conditions of the northern parts of the UK it would appear necessary to house weaned red deer calves in November (Blaxter *et al*, 1981; Hamilton and Blaxter, 1981).

A common practice at weaning is to house calves and train them to eat concentrate and other dry feeds over a period of 2-3 weeks. When housed later in the early winter, they accept such feeds readily, thereby reducing inappetance and stress. Calves show a period of winter inappetance between December and February, which is also associated with a period of low growth rate. In March, appetite increases by on average 25% and with it liveweight gain (Milne *et al*, 1978; Kay, 1979; Suttie and Simpson, 1985; Adam and Moir, 1985; Milne *et al*, 1987). Calves have been wintered successfully on a range of diets including hay, silage, and ammonia-treated straw as roughage sources, cereal-based concentrates and dried gras pellets(Adam and Moir, 1985; Blaxter *et al*, 1988).

Compensatory growth at pasture (25-50 g/day) following a period of low nutrition in the winter period has been observed by Adam and Moir (1985) and Milne *et al* (1987). Milne *et al* (1987) also demonstrated that compensatory growth at pasture occurred when a period of low nutrition, given in January and February, was followed by the feeding of a moderate quality diet given *ad libitum* between early March and early May before the calves were turned out to pasture. This pattern of winter feeding was associated with achieving the same liveweight in September as a high level of nutrition throughout the winter (see Figure 2) and is an efficient means of utilising expensive feed resources in the winter. No differences were observed in the carcase composition of animals exhibiting compensatory growth compared to those grown continuously when slaughtered at 16 months of age (Adam and Moir, 1985). At pasture liveweight gains of yearlings of 150-200 g/day are found with there being a linear relationship between sward height and liveweight gain up to a height of 8 cm (Milne *et al*, 1987).

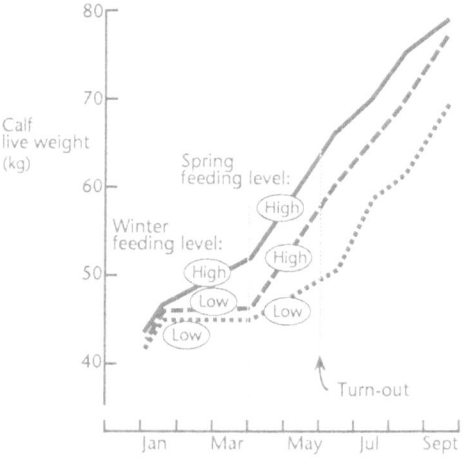

FIGURE 2. The effect of winter and spring patterns of growth of yearling stags on compensatory growth at pasture in the summer

The objective of current deer farming systems is the production of yearlings for slaughter at 16-18 months of age in the period from September to November at a liveweight of 80 kg, This is also a liveweight where there is a high probability of yearling females being successfully mated. These liveweights can be readily achieved with moderate growth rates during the winter followed by grazing swards at a height of 5-6 cm during the summer.

Slaughter in the UK is currently carried out by shooting in the field or by the use of existing red meat abbatoirs specially adapted for the purpose. A full description of the welfare and hygiene considerations of the two methods is given by Blaxter *et al* (1988). The long-term future of the deer industry in the UK is likely to be fulfilled by abbatoir slaughter, which is the method most widely used in New Zealand.

COMPARISONS WITH OTHER DOMESTICATED SPECIES

Red deer have been found to digest fibrous foods less completely than sheep. This is associated with a more rapid passage of undigested residues through the digestive tract. Red deer also eat more than sheep and, although there is a strong seasonal pattern in intake and metabolic rate, this is not associated with changes in passage time or digestive efficiency (see Kay *et al* (1984)). This would suggest that deer may have advantages over sheep in some circumstances. However the energy requirements for maintenance of liveweight of deer are higher than those of sheep, being more akin to those of cattle (see Kay (1984)). There is little evidence that the utilisation of energy for growth is different between species. Milne (1980) concluded that the differences so far identified between the two species in digestive physiology and metabolism are unlikely to lead to large differences in the biological efficiency with which they use pastoral resources. Differences in biological efficiency between deer and cattle or sheep are likely to be as a result of different dates of parturition, different fat contents of the body and differences in longevity and reproductive rate.

SYSTEMS OF PRODUCTION

Red deer performance and output from the use of hill resources in the UK will depend upon stocking rate and the long-term stability of the natural vegetation of such systems, which in turn depend upon the effect of grazing on sustainable plant productivity. On heather moorland, a common vegetation type in Scotland, stocking rates of 1 hind equivalent per hectare are consistent with the maintenance of the sward (Grant *et al*, 1981) and the provision of adequate nutrition from pasture. Systems in such areas have low inputs with the production of weaned calves weighing about 35 kg at 100 days old. The major limitation to their economic success is the high costs associated with fencing large areas of relatively poor quality vegetation.

On lowland areas output of weaned calf per unit area is considerably higher because of higher stocking rates (12-14 hinds per hectare) and more rapid growth rates of calves to weaning. Liveweights at weaning at 100 days of age are 40-45 kg. It is also easier, in lowland areas, to integrate finishing systems to systems of weaned calf production. Initial fencing costs are much lower than for hill systems but the cost of winter feeding, which has to be provided from conserved or purchased feed, is three times higher.

Systems based on a combination of upland sown swards grazed in the summer and an adjacent area of hill used to provide a cheap source of nutrients in the winter has considerable advantages and such systems are considered to be the most economically efficient at present.

Yearling stags for slaughter can be grazed at a stocking rate of 20 animals per hectare on sown grass swards in the UK. Antlers are removed once the antler has become hard, prior to the transfer of animals to an abbatoir for slaughter, Disbudding at 6 months of age has been found to be effective in halting the growth of antlers in yearling red deer (Hamilton, 1987). Purchased calves can be housed in buildings used previously for cattle in groups of up to 20 and additional fencing is only required for pastures for grazing yearlings in the summer. A review of the economics of systems of deer production in the UK is given by Hamilton (1986).

Marketing of home-produced deer-meat is in its infancy in countries such as the UK where deer meat is not commonly eaten. This will change as supplies become more available but a need to supply the market over a wider period of the year has been identified and current research is aimed at developing systems of production to enable this to occur. Two options being investigated are the use of melatonin to bring forward the date of birth of calves and thus allow calves to be slaughtered in their first year of life and the hybridisation of the red deer with larger species of deer to increase calf growth rates. Whilst there are several important management and disease issues which remain to be resolved, this review has indicated the large body of knowledge that already exists. The future of deer farming in Europe may be determined more by economic, fiscal and marketing issues than by a lack of understanding of the nutrition and management of deer.

REFERENCES

1. Adam CL and Atkinson T: Journal of Reproduction and Fertility, 72, 463-466, 1984.
2. Adam CL and Moir CE: Animal Production, 40, 135-141, 1985.
3. Adam CL, McDonald I, Moir CE and Pennie K: Animal Production, 46, 131-138, 1988.

4. Arman P, Kay RNB, Goodall GD and Sharman GAM: Journal of
    Reproduction and Fertility, 37, 67-84, 1974.
5. Blaxter KL and Hamilton WJ: Journal of Agricultural Science,
    Cambridge, 95, 275-284, 1980.
6. Blaxter KL, Boyne AW and Hamilton WJ: Journal of Agricultural
    Science, Cambridge, 96, 115-128, 1981.
7. Blaxter KL, Kay RNB, Sharman GAM, Cunningham JMM and Hamilton
    WJ: Farming the red deer. Her Majesty's Stationery Office,
    Edinburgh, 1974.
8. Blaxter KL, Kay RNB, Sharman GAM, Cunningham JMM, Eadie J and
    Hamilton WJ: Farming the Red Deer. Her Majesty's Stationery
    Office, Edinburgh, 1988.
9. Grant SA, Hamilton WJ and Souter C: Journal of Ecology, 69,
    189-204, 1981.
10. Hamilton WJ: RASE Reference Book, Royal Agricultural Society of
    England, Kenilworth pp 165-168, 1986.
11. Hamilton WJ: Macaulay Land Use Research Institute Annual Report
    (in press) 1987.
12. Hamilton WJ and Blaxter KL: Journal of Agricultural Science,
    Cambridge, 95, 261-273, 1980.
13. Hamilton WJ and Blaxter KL: Journal of Agricultural Science,
    Cambridge, 96, 115-128, 1981.
14. Henderson, TG: New Zealand Veterinary Journal, 31, 221-224,
    1983.
15. Kay RNB: Agricultural Research Council Research Review, 5,
    13-15, 1979.
16. Kay RNB: In The Biology of Deer Production (ed P Fenessy and KR
    Drew) Wellington, Royal Society of New Zealand pp 411-422, 1984.
17. Kay RNB, Milne JA and Hamilton WJ: Nutrition of red deer for
    meat production. Proceedings of the Royal Society of Edinburgh,
    82B, 231-242, 1984.
18. Loudon ASI, Dorroch AD and Milne JA: Journal of Agricultural
    Science, Cambridge, 102, 158-167, 1984.
19. Loudon ASI, McNeilly AS and Milne JA: Nature, 302, 145-147,
    1983.
20. Milne, JA: Proceedings of the New Zealand Society of Animal
    Production, 40, 151-157, 1980.
21. Milne JA, Russel AJF and Hamilton WJ: Macaulay Land Use
    Research Institute Annual Report (in press) 1987.
22. Milne JA, MacRae JC, Spence AM and Wilson S: British Journal of
    Nutrition, 40, 347-357, 1978.
23. Milne JA, Sibbald, Angela M, McCormack Heather A and Loudon ASI:
    Animal Production, 45, 511-522, 1987.

<u>SESSION I</u>

Chairman: Dr. R.J. Jorgensen

Co-Chairman: Dr. R. Munro

STUDIES ON THE EPIDEMIOLOGICAL PATTERN AND CONTROL
OF NEMATODE INFECTION IN CERVIDAE

F.H.M. Borgsteede
Central Veterinary Institute
P.O. Box 65, 8200 AB Lelystad
The Netherlands

ABSTRACT

The pattern of gastrointestinal helminth egg output in reindeer kept under semi-natural conditions in the Netherlands was studied from 1976 to 1987. In the first six years the animals were treated every two months and thereafter with longer intervals with a benzimidazole drug mixed into their additional food. Post-mortem worm counts were carried out on animals that had died or were eliminated from the herd. Four age groups were distinguished: < 1 year, 1-2 years, 2-3 years and > 3 years. The pattern of trichostrongyle egg output was similar in all age groups. The first peak was observed in autumn and the second in spring. The egg output was considerably higher in animals under 1 year of age than in other groups, indicating that immunity had developed. The first peak was thought to reflect infection picked up during the summer months (Type I), the second was possibly the result of maturation of inhibited larvae (Type II). Treatment produced limited results on these patterns. *Trichuris* eggs were seen almost exclusively in first grazing season animals. *Capillaria* eggs were also observed in older age groups. A light infection with *Dictyocaulus viviparus* was always present in the herd, but without clinical signs. Worm counts in 30 reindeer varied from 1,500 -275,200. *Ostertagia leptospicularis* was most common (90.9 %), followed by *Skrjabinagia kolchida* (7.6 %) and *Spiculopteragia boehmi* (1.5%). The percentage of arrested larvae was highest in the winter months with peaks in December and February of more than 90%.

INTRODUCTION

The gastrointestinal nematode fauna of wild Cervidae has been well described in many European countries. In the Netherlands, for instance, Swierstra et al. (1959) and Jansen (1963) have listed the parasites of roe deer, red deer and fallow deer. Long term studies on the epidemiological pattern of the helminths in these animals have proven problematic compared with studies in ruminants such as cattle and sheep. Not surprisingly, most relevant knowledge of the epidemiology of helminth infections in Cervidae has been derived from reindeer (*Rangifer tarandus*), the only domesticated Cervid in Europe. Particularly in Scandinavian countries much work on this subject has been done (Christensson and Rehbinder,1975; Rehbinder and Christensson,1977; Rehbinder and von Szokolay,1978; Halvorsen, 1986). Leader-Williams (1980) observed the internal parasites of Norwegian reindeer introduced to the sub-antarctic island of South Georgia. Bye (1986) described the abomasal nematodes of three Norwegian wild reindeer populations.

Knowledge of the influence of anthelmintic treatment on the epidemiological pattern is almost nonexistent. Reindeer herds are generally not treated systematically (Nordkvist, 1971). In contrast with farmed deer the effect of treating wild deer by mixing anthelmintics in food has not been systematically followed over a long period (Düwel et al. 1979; Kutzer and Prosl, 1979; Mason and Gladden, 1983).

In the present study results are described of a ten year study of a reindeer herd kept under semi-natural conditions in the Netherlands.

MATERIALS AND METHODS
Animals
The study was carried out in "Nature Park Lelystad", where reindeer, elk, Father Davids deer and European bisons are kept under semi-natural conditions. Reindeer were introduced into the park in 1975, and by the time the present study began in November 1976, the herd consisted of 20 animals. Table 1 shows the structure of the herd during the period 1976 - 1987.

The reindeer herd was kept on an area of 30 ha. Ten ha. consisted of grassland that was mown twice each summer. The mown vegeation was not removed. Another ten ha. of grassland was mown once each year, but all mown vegetable material was removed. The last ten ha. were not covered with grass, but trees had been planted. With the exception of some elder bushes this area was totally destroyed by the animals. No cultivation was carried out on that portion, and in summer the animals did not graze there. Beginning in July 1985 the total area was divided into two equal parts. Grazing on each part was allowed for one year, after which the animals were moved to the other part. Daily 17 kg. of additional food was given (sheep concentrates, flattened oak, wheat bran and dry sugar-beet pulp), together with some willow twigs. On severe winter days, when the ground was covered with snow, more food was supplied.

TABLE 1   Number of reindeer per year.

Year	April	July	Born	Died
1976	End of the year : 20 animals			
1977	20	27	7	3
1978	24	31	7	3
1979	28	34	6	3
1980	31	37	6	7
1981	30	36	6	4
1982	32	36	4	3
1983	33	39	6	7
1984	32	39	7	12
1985	27	35	8	8
1986	27	33	6	0
1987	33	40	7	-

Treatments

Beginning in March 1977 the reindeer were treated every two months with fenbendazole mixed into the additional food. The total amount of the drug used was based on the estimated weight of the herd in kg. and a dose rate of 5 mg/kg body weight. From September 1979 on the treatment regime was changed: on two consecutive days a dose of 5mg/kg was given. From December 1983 on fenbendazole was replaced by oxfendazole. Beginning in 1985 young calves were individually treated in autumn with an injection of 0.2 mg/kg ivermectin.

Parameters

Every four weeks faecal samples of each age group were collected when freshly deposited. Eggs were counted with a modified McMaster technique (Borgsteede and Hendriks, 1973) and infective larvae were cultured and identified (Borgsteede and Hendriks, 1974). Post-mortem worm counts were made from the contents of the abomasa and small intestines of animals that had died naturally or that for other reasons were eliminated from the herd. The contents were diluted to 10 L and samples of 1/100 were collected, sieved over a screen with 74 μm mesh and preserved in 4% saline-formalin solution for counting and identification.

RESULTS
Trichostrongyle egg output

Almost all of the excreted eggs were of the "strongyle" type. The results of the examinations of the different age groups were averaged over the years and the patterns summarized in figs. 1-4. In Fig. 1,3 and 4 two peaks can be distinguished : first in the period September to November, second in February to April. In Fig. 2 only the second peak is obvious. Egg output in first grazing season calves was much higher than in the other three groups. No differences could be found between these groups. Influence of treatment on these patterns appeared limited. After culturing the faeces only one type of larva, belonging to the *Ostertagia*-group, was found.

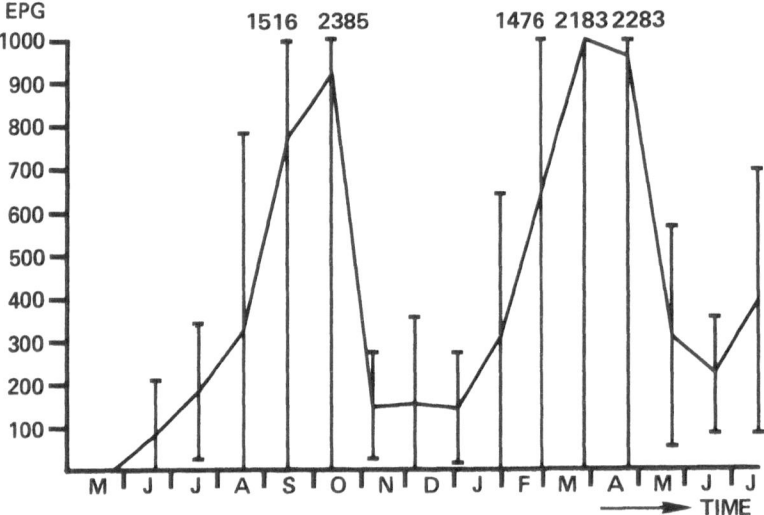

Fig. 1 Average EPG-pattern (± S.D.) of reindeer less than 1 year old over the period 1977-1987.

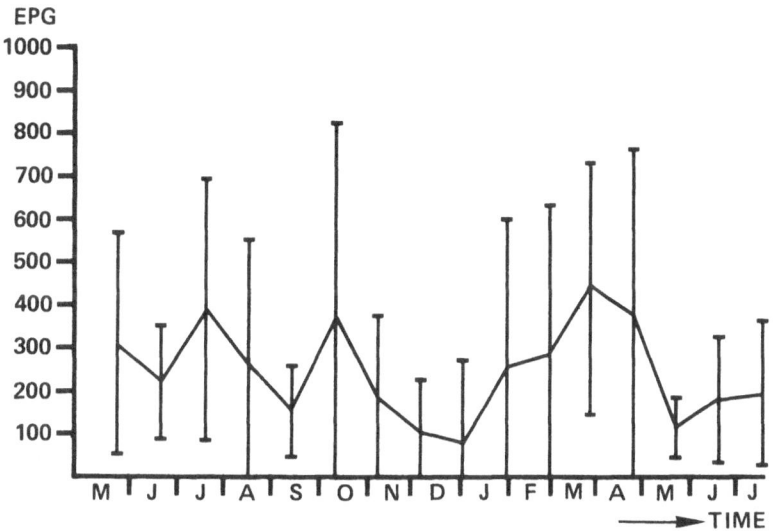

Fig. 2 Average EPG-pattern (± S.D.) of reindeer 1-2 years old over the period 1977-1987.

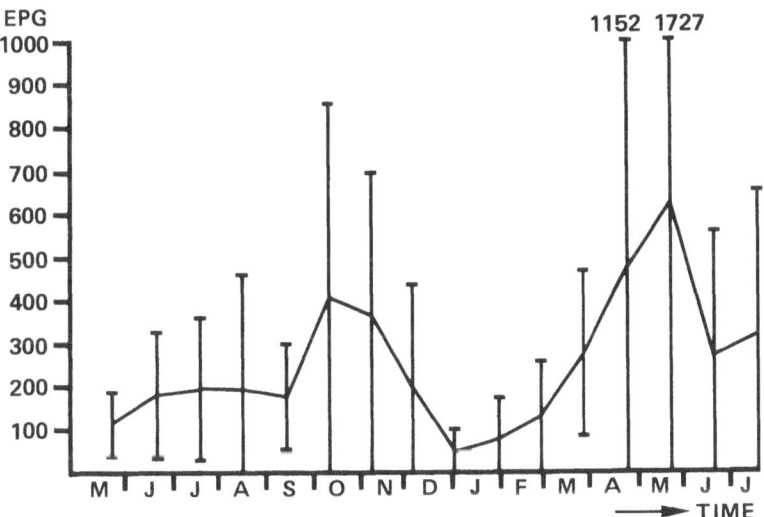

Fig. 3 Average EPG-pattern (± S.D.) of reindeer 2-3 years old over the period 1977-1987

18

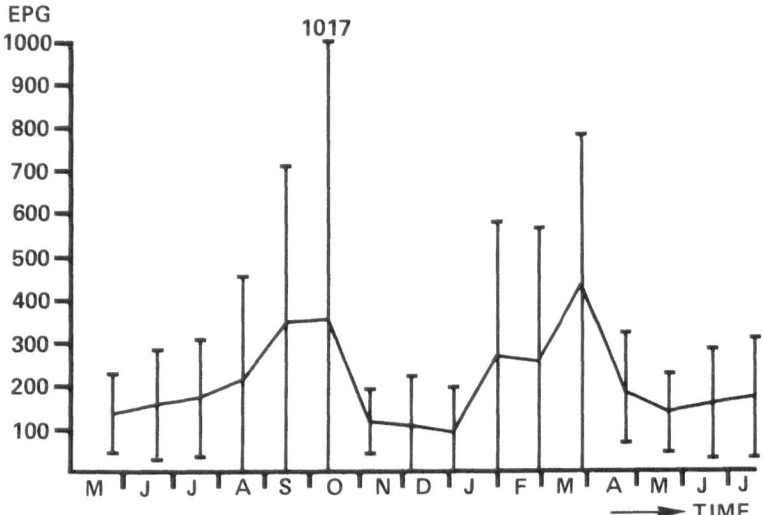

Fig. 4 Average EPG-pattern (± S.D.) of reindeer older than 3 years over the periode 1977-1987.

Output of other eggs and of larvae

Frequently in first grazing season animals low numbers of *Trichuris* eggs were found during the period August-December.

Thereafter, this egg type was seen only incidentally. In some years *Capillaria* eggs were observed in calves from November onwards, but always in very low numbers. In older age groups *Capillaria* eggs were seen regularly. During the period of investigation *Nematodirus* eggs were found only twice, in January and in February 1986, in calves born in 1985.

Although not systematically followed, lungworm larvae (*Dictyocaulus viviparus*) were observed from time to time in very low numbers. Larvae of *Elaphostrongylus rangiferi* were never seen.

Post-mortem worm counts

The nematode parasites found in this herd are given in Table 2.

TABLE 2   Nematode species found in reindeer over the
                            period 1976-1987.

Lungs	*Dictyocaulus viviparus*
Abomasum	*Ostertagia leptospicularis*
	Skrjabinagia kolchida
	Spiculopteragia boehmi
	Haemonchus contortus
Small intestine	*Capillaria sp.*
Colon and caecum	*Trichuris sp.*

Haemonchus contortus was found only once in an animal that had
died just before the study started. Thereafter, a regular treatment
was given and *H.contortus* was never observed again. *D.viviparus* was
present in low numbers (always less than 10) in most of the examined
animals, but never gave rise to serious problems. The gastrointestinal
worm burden was counted in 30 animals. The parasites found were almost
exclusively abomasal. In the samples of the small intestine no worms
were found. In some reindeer the whole intestinal contents were
examined, but only a few *Capillaria* sp. were found. The mean abomasal
worm burden was 65,000, ranging from 1,500-275,200. By far the most
common nematode found was *O.leptospicularis* (90.9 %), followed by
Skrjabinagia kolchida (7.6 %) and *Spiculopteragia boehmi* (1.5 %) The
average percentage of arrested stages was 37, ranging from 0-95.8. The
percentage of arrested larvae was highest in the winter months with
peaks in December and February of more than 90.

DISCUSSION

   The results of this study proved interesting in several respects.
We found that within and between age groups over the years patterns of
trichostrongyle egg ouput were highly consistent. Two peaks were
found, resembling the pattern of *Ostertagia ostertagi* infections in
young calves described by Anderson et al. (1965). The latter described
two types of disease, one occuring in autumn (Type I), the other in
winter (Type II). The first type resulted after ingestion of infective
larvae from the undisturbed development of adult worms, while Type II
occurred when massive numbers of arrested larvae matured. The

post-mortem worm counts of our study confirmed these findings. Although the number of reindeer investigated was rather low for drawing conclusions about worm numbers throughout the year, there was a strong indication that the percentage of arrested larve was highest in the winter months. No data on the number or the percentage of arrested larvae could be found in the literature, although Rehbinder and Christensson (1977) suggested the possibility of winterostertagiasis (=Type II) in reindeer calves in their study. Compared with other studies the worm numbers we found were rather high.Rehbinder and Christensson (1977) found low numbers in 26 autumn-slaughtered bulls without indicating the exact worm burden. Rehbinder and von Szokolay (1978) found on the average less than 100 (2,300 max.) in 40 calves and 900 (2,200 max.) in 14 older animals, all slaughtered during the period of December to March. The occurrence of Type II osteragiasis in our study could be ascribed to the weakened immunity of the calves during the winter months. Regular treatment had hardly any influence, which may be explained by the lower intake of additional food in which the anthelmintic was mixed by the more heavily parasitized animals. On the other hand the egg output of animals in their first grazing season was considerably higher than in the other groups, indicating a certain degree of developed immunity.

The patterns of *Trichuris* egg output were also fully consistent with those of farm animals such as cattle and sheep. Eggs were mainly observed in the first grazing season. The first positive samples were from August. Calves are born in May, thus this is consistent with the intake of infective eggs just after birth combined with a tree month prepatent period. In older age groups *Trichuris* eggs were rare, indicating the development of immunity after the first infection. This is in sharp contrast with elks (*Alces alces*), kept on the same grounds under the same conditions, which appeared to build up no immunity at all.

Capillaria infections also occurred in older reindeer, but like *Trichuris* probably caused no trouble. In the future the number of *Trichuris* and *Capillaria* may possibly increase, since the dose of anthelmintic now given to the animals is assumed to be too low to kill these parasites.

Another problem that may rise from the use of an intensive treatment schedule is the development of anthelmintic resistance. In order to determine any existing drug resistance, an *in vitro* egg development test were carried out in 1986. The LC50 ranged between 0.02 and 0.05 ppm thiabendazole. These values are regarded as not indicative for resistance, although for a good comparison a "susceptible" strain of *O.leptospicularis* is required. The improbability of drug resistance is confirmed by the reflection that the vast majority of the parasite population is not under anthelmintic pressure, but on the pasture (the larvae are in so-called refugia). On the other hand, possible underdosing of the herd could enhance the increase of the presence of resistant genes, if present in the worm population.

The important lesson learned from our experiences with the park and its inhabitants is that a parasite free start in an uncontaminated environment is very necessary to avoid problems. If individual animals or herds are brought into a new area they should be treated with an overdose of a modern broad spectrum anthelmintic, kept 48 hours in quarantine to be sure that the faeces do not contain viable eggs, and then turned out into clean pasture or a clean area wich has been free of other grazing ruminants for at least one year. If these rules are followed, helminth parasites will be no problem and treatment will be unnecessary. The proof is a herd of Father Davids deer, introduced into the park in 1979, but kept in a separate area. They were turned out free of helminths and up till now no eggs have been seen in the faeces, nor has it been possible to culture larvae.

ACKNOWLEDGEMENTS

The author wishes to acknowledge the assistance given by the co-workers of the Nature Park Lelystad and Dr. J.R. Hoedemaker. Within the Parasitology Department much work was done by Mr. W. van der Burg, and Mr. R. Mes. Dr. J. Jansen kindly confirmed the identification of the adult worms. Mrs. V. Thatcher corrected the grammatical mistakes of the first draft of this manuscript.

REFERENCES

Anderson, N., Armour, J., Jarret, W.H.F., Jennings, F.W., Ritchie, J.S.D. and Urquhart, G.M. 1965. A field study of parasitic gastritis in cattle. Vec.Rec., 77,1196-1204.

Borgsteede, F.H.M. and Hendriks, J. 1973. Een kwantitatieve methode voor het kweken en verzamelen van infectieuze larven van maagdarmwormen. Tijdschr.Diergeneesk.,98,280-286.

Borgsteede, F.H.M. and Hendriks, J. 1974. Identification of infective larvae of gastrointestinal nematodes in cattle. Tijdschr. Diergeneesk.,99,103-113.

Bye, K. 1986. Abomasal nematodes from three Norwegian wild reindeer populations. Can. J.Zool., 65,677-680.

Christensson, D. and Rehbinder, C. 1975. Parasites in reindeer calves-faeces examination. Nord.Vet.-Med.,27,496-498.

Düwel, D.,Kirch,R. und Tiefenbach,B. 1979. Zur Behandlung des Nematoden-Befalls beim Wild mit Panacur. Berl. Münch.Tierärztl.Wschr.,92,334-339.

Halvorsen,O. 1986. Epidemiology of reindeer parasites. Parasitol.Today, 2,334-339.

Jansen Jr.,J. 1963. Some problems related to the parasite interrelationship of deer and domestic animals. Int. Union of Game Biologists. Transactions VI th Congr., 127-131.

Kutzer,E. und Prosl,H 1979. Zur anthelmintischen Wirkung von Fenbendazaol (Panacur^R) bei Rothirsch (Cervus elaphus hippelaphus) und Wildschwein (Sus scrofa). Wien tierärztl. Mschr., 66,285-290.

Leader-Williams, N. 1980. Observations on the internal parasites of reindeer introduced into South Georgia.Vet.Rec., 107,393-395.

Mason,P.C. and Gladden, N.R. 1983. Survey of internal parasitism and anthelmintic use in farmed deer. N.Z.Vet.J.,31,217-220.

Nordkvist, M. 1971. Über die Probleme der Veterinarmedizin in der Rentierzucht. Vet.Med.Nachr.,2/3,397-405.

Rehbinder, C. and Christensson,D. 1977. The presence of gastro-intestinal parasites in autumn-slaughtered reindeer bulls. Nord.Vet.-Med.,29,556-557.

Rehbinder, C. and von Szokolay,P. 1978. The presence of gastric parasites in winter-slaughtered reindeer. Nord.Vet.-Med,30,214-216.

Swierstra, D., Jansen Jr. J. and Van den Broek,E. 1959. Parasites of animals in the Netherlands. Tijdschr. Diergeneesk.,84,892-900.

EPIDEMIOLOGY AND CONTROL OF PARASITIC
DISEASES IN DANISH DEER FARMS

R.J. Jorgensen*, F. Vigh-Larsen**
*Institute of Internal Medicine, 13 Bulowsvej,
DK-1870 Frederiksberg C, Denmark
** National Institute of Animal Science, Forsogsanlaeg, Foulum,
Postboks 39, DK-8833, Orum Sdr. Lyng, Denmark

Faecal analysis was performed on five Danish deer farms at regular intervals in order to follow the level of infection with internal parasites. The details of two farms are reported below in a summarized form. Farm A was grazing approximately 35 fallow (Dama dama) with calves on a 4 hectare permanent pasture. Winter feeding consisted of hay fed in racks and whole oat in troughs. Gastrointestinal parasitism culminated in the period January-May among calves born the previous summer. Predominant symptoms were diarrhoea and loss of ability to follow the herd. Closer examinations revealed emaciation. The collapse of these calves was seen as the combined effect of 1. gastrointestinal parasitism, 2. insufficient uptake of grain by these lower ranking animals due to competition and 3. increased loss of bodywarmth due to incomplete winter coat caused by scratching in response to the presence of numerous lice (Damalinia sp.). Lungworm (Dictyocaulus viviparus) larval counts increased in late summer up to 200 L/10 gram among the calves in September at which time the flock was treated. Group dosing with Albendazole (Valbazen R vet.) at 5g per ton daily for 4 days eliminated faecal excretion of larvae and eggs.

Farm B was new and composed of hinds imported from U.K. and B.R.D. Elaphostrongylus cervi larvae were diagnosed in faeces from hinds of German origin (Jorgensen and Vigh-Larsen, 1986), whereas Mycobacterium paratuberculosis was first diagnosed in hinds from U.K. (Jorgensen and Berg-Jorgensen, 1987). The major parasitic problem was caused by Dictyocaulus viviparus during August-September among calves and in particular in yearlings which had been housed and which had excreted large numbers of larvae before they were treated with Ivomec and housed the previous autumn. Clinical symptoms were coughing and weight loss. M paratuberculosis was cultured from several faecal samples and caused the death of one hind. Guidelines for the control of internal parasites are difficult to make due to large farm-to-farm variations in management. (Most farms are still grazing all age groups together in one set-stocked flock which is outwintered). In general it is recommended that deer should be given anthelmintic treatment:

1. at the start and at the end of the winter feeding period
2. at weaning
3. when moved to clean pasture
4. further treatments on indication (problem farms, high stocking rate)

Remember:

that treatment only has a short effect once the pasture is contaminated

that flock moving plus treatment is far more effective than treatment alone

that calves and yearlings are the groups at risk

Conclusions

1. Lungworms are present on some red deer farms but not on others
2. Lungworm problems may be severe in red deer calves following re-exposure after a period of no exposure (e.g. after housing)
3. Lungworms are most serious in red deer in the summer (periods of risk)
4. Lungworms affect red deer yearlings and calves
5. Gastrointestinal parasitism in fallow deer are most serious in late winter among calves
6. Separate winter-feeding of fallow deer calves with concentrate is essential when such calves are to be outwintered under Danish conditions
7. In supervision of parasitic infections by examination of fresh faeces collected from the ground, special attention should be paid to the collection of faeces from young animals and of soft or pasty faeces.
8. Maximum individual faecal counts of eggs and larvae is a better indication of the necessity for anthelmintic treatment than average counts

9.   Strongyle egg counts are a better indicator than total egg counts

10. Trichuris and Capillaria may reach high levels in outwintered fallow deer fed on the ground.

REFERENCES

Jorgensen, R.J. & Vigh-Larsen, F.   (1986).   Preliminary observations on lungworms in farmed and fereal red deer (<u>Cervus elaphus</u>) in Denmark. Nord. Vet.Med., <u>38</u>, 173-179

Jorgensen, R.J., & Berg-Jorgensen, J. (1987). paratuberkulose hos krondyr. (Paratuberculosis in red deer). Danks. Vet. Tidsskr, <u>70</u>, 322-324.

PULMONARY PARASITES: PATHOLOGY AND CONTROL

R. Munro

Moredun Research Institute, 408 Gilmerton Road,
Edinburgh EH17 7JH

ABSTRACT

Scottish red deer appear relatively tolerant to infestation by
Dictyocaulus and the pulmonary response to migrating and developing
Dictyocaulus larvae is considerably less severe than might be expected in
cattle. Heavy challenge may, however, result in loss of condition, poor
growth rates and unexpected deaths. The clinical signs, postmortem
appearance and histopathological findings are described and methods of
control are suggested. In a survey, lung lesions associated with the
eggs and larvae of Elaphostrongylus cervi were found in approximately 76%
of adult free-living red deer in Scotland. In 14% of these cases a
granulomatous pneumonia was present. The mean infection rate in
diagnostic samples from farmed red deer, over 6 months of age, between
1977-84, was 50.2%. None of the farmed animals showed significant
pneumonia.

INTRODUCTION

Pulmonary pathology is a poorly studied aspect of deer disease and
there are few descriptions of histopathological changes in red deer lung.
Of the pulmonary conditions studied, lungworm infestation has attracted
the greatest interest. This paper outlines the changes caused by these
parasites and suggests means of control.

DICTYOCAULUS INFECTON

Dictyocaulus has long been recognised as a parasite of Scottish deer
and D. viviparus was the only species found by Cameron (1932) in
Scotland. D. viviparus has also been reported in Cervus elaphus from
many European countries, North America and New Zealand. Other species
of Dictyocaulus, namely D. eckerti and D. hadweni, have been
described in red deer and its close relatives, marals and wapiti. These
three parasites may be synonymous although not all taxonomists are
convinced. Boev (1963) could find no morphological differences between D.
vivaparus and D. eckerti but was of the opinion that they are
"biological species", D. viviparus occurring in cattle and D. eckerti

27

in deer. Dunn (1967) cautiously refers to Dictyocaulus sp. in British deer but suggests that "deer have at least a separate strain and probably a separate species."

Clinical and Post-Mortem Descriptions of Dictyocaulus infection

In the early days of deer farming no particular clinical effects were noted to be associated with D. viviparus infestation (Blaxter et al., 1974). However the prediction that under intensive conditions parasitic pneumonia might be a problem (McDiarmid, 1969) has since proved to be the case. In 1975, gradual loss of weight and unthriftness was ascribed to D. viviparus infestation at Glensaugh (Corrigall et al., 1980). New Zealand deer farms suffered similarly, with Dictyocaulus described as the major "problem" parasite and many deaths being attributed to heavy infection whilst lesser burdens caused loss of productivity (Mason, 1979). Besides the rather non-specific signs of loss of condition, retarded growth rates and roughened coats, unexpected deaths may occur and continue over a period of weeks (Wilson, 1979; Charleston, 1980; Reenan, 1982). Clinical signs in heavy experimental infections are described by Corrigall and colleagues (1982). The most severely affected animals showed tachypnoea, dyspnoea and a variety of harsh sounds on auscultation accompanied by increased heart rate, some coughing and decreased appetite. These effects were most apparent during the third and fourth weeks after initial challenge. There is general agreement that young stock tend to be more heavily parasitized than mature animals. Experience in Czechoslovakia has shown that nutritional factors are very important in Dictyocauliasis and animals weakened by starvation are particularly susceptible to infection (Erhardova-Kotrla and Kotrly, 1973).

Descriptions of the damage caused to the lungs of red deer by natural Dictyocaulus infections are limited. Light to moderate burdens, i.e. up to 20 worms, cause little or no gross changes (Volkholz, 1974; Wilson, 1979). In some animals with heavier infections dark red slightly sunken areas of pulmonary consolidation are present (Corrigall et al., 1980); in others the lungs are hyperaemic or fail to collapse on opening the thorax (Wilson, 1979; Charleston, 1980). Lung worms in frothy exudate can be found throughout the bronchial tree and, in sufficient numbers, may block the bronchi and bronchioles.

Histopathology of experimental Dictyocaulus infection

The following description is confined to lesions found in red deer calves, previously unexposed to Dictyocaulus, after experimental challenge with 500 third stage larvae per kg liveweight per os daily for 17 days. Fuller accounts of these experiments, lesions in vaccinated deer and in deer challenged with "deer-derived" larvae are given by Munro (1985).

Eighteen days post-challenge, heavy parasitic burdens were present in the bronchi and larger bronchioles. Certain areas were severely ulcerated with eosinophils and round cells migrating through the mucosa into the lumen (Figure 1). Much of the remaining mucosa of these airways

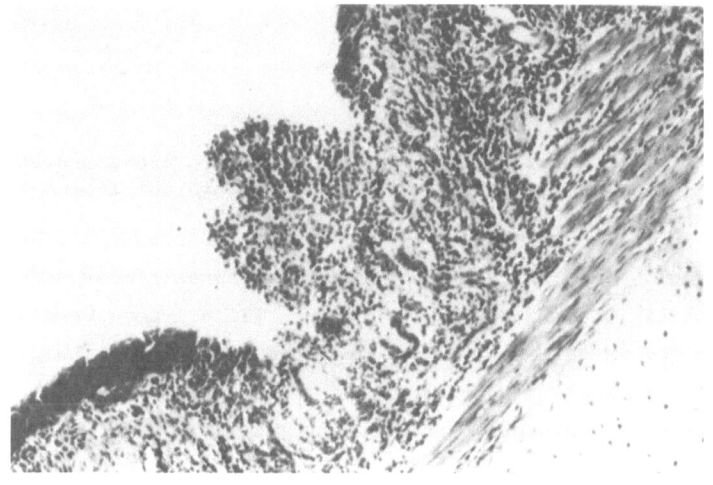

Fig. 1   Ulceration of the bronchial epithelium and migration
of inflammatory cells into the lumen.   HE x 190

was eroded and distorted. Peribronchial cuffing by large numbers of eosinophils, lymphocytes and some plasma cells was a prominent feature. Most of the small bronchioles were plugged by eosinophilic debris and degenerating epithelial cells. An uneven interstitial reaction included alveolar septal congestion, proliferating septal cells and an increase in round cells and eosinophils. Collapse of alveoli was confined to small areas adjacent to the airways (Figure 2). Within the lumina of

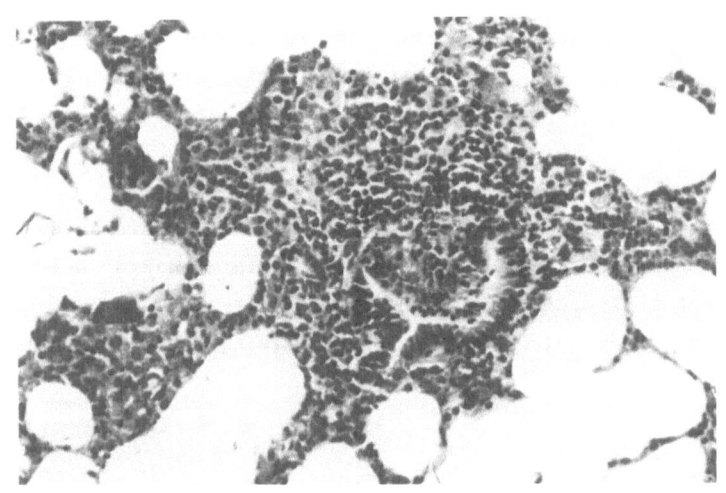

Fig. 2. Alveolar collapse is restricted to areas around the airways. Alveolar macrophages are common in the lumina of the partially collapsed alveoli. HE x 190

these alveoli were excess alveolar macrophages (many showing foamy vacuolation of the cytoplasm), oedema fluid, some swollen alveolar epithelium and an early neutrophil response. The interlobular septa were oedematous in small sections and contained moderate to heavy infiltrations by eosinophils and round cells.

Occlusive bronchitis and bronchiolitis were commonly found by day 24. The mucosa of these airways was necrotic and purulent exudate, sometimes containing recognisable parasitic remnants, filled the lumina. Cuffs of mononuclear cells, eosinophils and some neutrophils collected around the airways. At times these peribronchial and peribronchiolar tissues were oedematous. An alveolar response developed, for the most part, in restricted areas around the air passages. Cuboidal cells often replaced the normal epithelium in these alveoli and macrophages, neutrophils and some eosinophils filled the lumina (Figure 3).

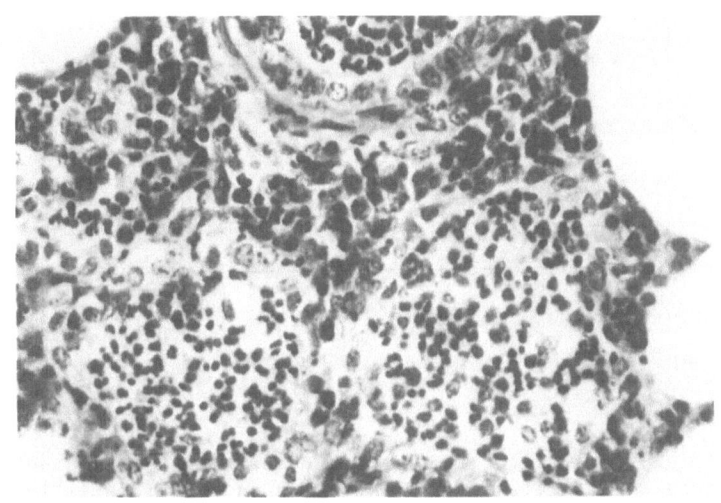

Fig. 3   Some alveoli adjacent to the airways are lined by cuboidal epithelium and collections of alveolar macrophages, neutrophils and eosinophils fill the lumina.   HE x 380

Thickening of the alveolar septa by round cells and engorgement of the capillaries added to the consolidation. In certain lobules there was a massive exudative response. This consisted of widespread odematous flooding, leakage of fibrin and collection of large numbers of alveolar macrophage and neutrophils in the alveolar lumina (Figure 4). The purulent reaction led, in places, to the formation of discrete abscesses. Within these "exudative" lobules parasitic larvae were numerous. The interlobular reaction varied; in some areas there was no change, some septa were oedematous whilst in other sections mononuclear or eosinophil infiltration with fibroplasia was the rule.

Ninety-five days after inoculation stenosis of the bronchioles and bronchi by well-developed lymphoreticular follicles, some of which had multiple germinal centres, could be found. The epithelium of the airways was intact but globule leuocyctes were seen between the epithelial cells

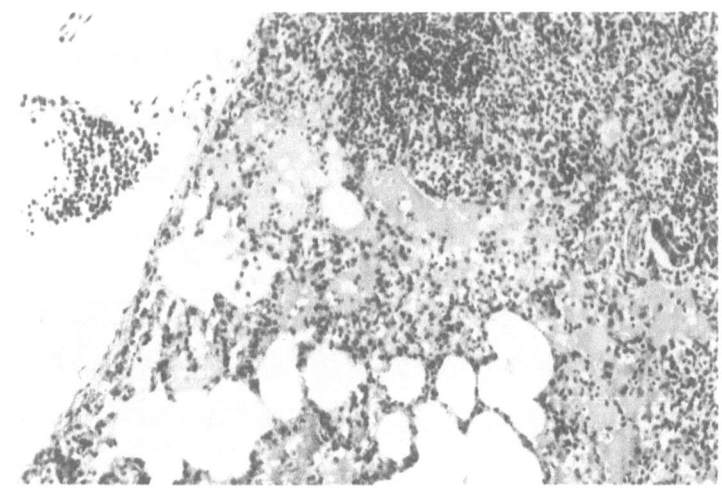

Fig. 4   An "exudative" lobule showing oedematous flooding, leakage
of fibrin and collection of macrophages and neutrophils in the
alveolar lumina.   HE x 114

and in the sub-epithelial tissue of the bronchi and larger bronchioles.
The alveolar septa of some animals were diffusely thickened by a
low-grade mononuclear hypercellularity. Lymphoid follicles, occasionally
multiple,   had formed in the interlobular septa which were themselves
irregularly thickened by fibroplasia and contained modest numbers of
eosinophils. No parasites were found at this stage.

Histopathology of natural Dictyocaulus infection

Typical severe responses show irregular ulceration of the bronchial
and bronchiolar epithelium.   Eosinophil infiltration is moderate in the
bronchiolar epithelium, often fairly marked in the lamina propria of the
airways and large collections form in the bronchial lumina. Medium or
large lymphoid follicles are present in the peri-bronchial and
peri-bronchiolar tissue and moderate collections of eosinophils are
distributed around the airways. The alveolar septa adjacent to damaged
airways or interlobular septa are hypercellular and contain excess
eosinophils.   Alveolar collapse is usually slight, some alveoli contain

an excess number of alveolar macrophages and those alveoli adjacent to the airways contain moderate numbers of eosinophils. Certain lobules have a marked exudative response consisting of intra-alveolar oedema and haemorrhage usually accompanied by excess macrophages and eosinophils. These "exudative lobules" are surrounded by intra-lobular septa greatly thickened by oedema, fibroplasia and heavy infiltrations of eosinophils and mononuclear cells. Generally, eosinophil infiltration of the intra-lobular septa in these severely affected lungs is moderate but patchily distributed. Lymphoid follicle formations and collections of lymphocytes are common in the intra-lobular septa and sub-pleural connective tissue.

Many lungs have lesser reactions. In these the epithelium of the airways is intact and the eosinophil infiltration of the walls of the bronchi and bronchioles is often absent or restricted to moderate numbers and occur only in the lamina propria. Peri-bronchiolar eosinophil numbers are frequently higher than around the larger air passages and similarly, lymphoid follicles are often more developed adjacent to the bronchioles (Figure 5). Eosinophil and macrophage numbers in the alveoli

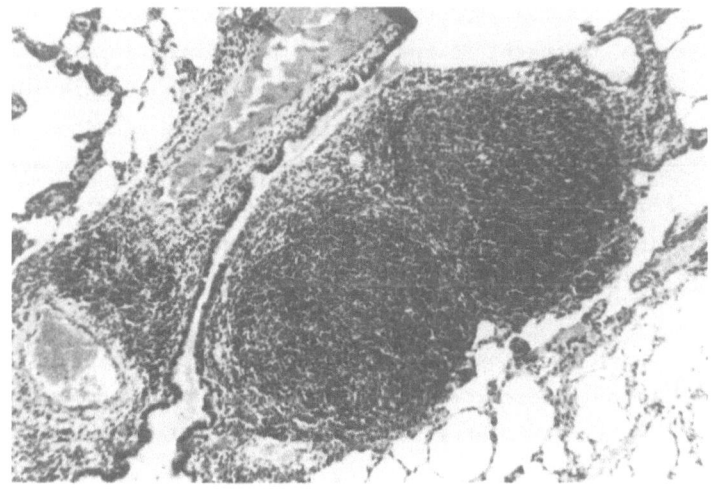

Fig. 5   Compression of a bronchiole by lymphoid follicle formation.   HE x 76

are generally low but foci occur regularly. Inter-lobular and sub-pleural reactions vary greatly from those showing no reaction to others with focal collections of eosinophils and round cells and limited fibrosis.

Control of Dictyocaulus infection

The work of Corrigall et al (1982) showed that deer appear capable of withstanding levels of challenge which would have proved severe if not fatal in cattle. A partial explanation for the differing clinical pictures between cattle and deer stems from the remarkable tolerance of red deer to the passage of larvae through the alveoli. In deer the bronchial damage is somewhat similar to that which occurs in cattle but the pneumonia associated with aspirated eggs and larvae was not extensive and alveolar epithelialisation was restricted to relatively small areas. As a result, the "vicious circle" of lungworm infection described for cattle, whereby consolidation, collapse, epithelialisation and acute bronchial damage combine to cause anoxia with subsequent cardiac impairment leading to pulmonary oedema, was not found in infected deer. In the post-patent phase of the disease in cattle, alveolar epithelialisation remains a major cause of respiratory embarrassment. In deer, however, there is no tendency towards widespread alveolar epithelialisation between 50 and 95 days post infection. Consequently post-patent husk does not appear to be a problem in experimentally or naturally infected deer. The implications of these findings, as far as control measures are concerned, are that efforts, in deer, can be concentrated on achieving a balance between the number of parasites required to stimulate natural immunity and ensuring that the parasitic burden is below that which reduces productivity.

Animals which have not had the opportunity to develop a degree of immunity are at the greatest risk. Hinds off the hill and yearlings (and calves in farms in the wetter and warmer parts of the country) are the most susceptible groups. Cost efficient control depends on formulating farm-specific programmes and on strategic anthelmintic treatment aimed primarily at the high risk categories. Farm-specific programmes will include

1. Monitoring of faeces for larvae by the Baermann technique, by collecting 10 freshly passed samples from calves, yearlings and adults in early, middle and late summer and in the autumn.

2. Keeping good records of weight gains.

3. Resampling any groups of animals coughing or showing loss of condition.

4. Using strategic dosing. (Treat only those animals which require to be treated). Adults should be treated in early summer to reduce pasture contamination before calving. It may then not be necessary to treat the adults again but this will depend on the level of challenge and stocking density. In yearlings, faecal larval counts correlate well with the number of adult parasites in the airways and so they should be re-treated as often as faecal sampling and the weight records indicate that they are carrying a significant burden. Regular treatment at 3-5 weeks intervals throughout the summer is expensive and may not be necessary. Conditions vary greatly from one part of the country to the next, the drier east having less of a problem than the west.

5. Treating calves at weaning or when being housed for the first time. (On some farms calves require to be treated before weaning).

Table 1 lists the products and dosage rages which have been shown

TABLE 1   Dictyocaulus : Anthelmintic Treatment

Dosing - Oxfendazole	4.5 mg/kg/BW
Fenbendazole	7.5 mg/kg/BW
Febantel	7.5 mg/kg/BW
Albendazole	10 mg/kg/BW
Injection - Ivermectin	200 mcg/kg/BW
In-feed - Albendazole	100 mg/head/10 days

(under experimental conditions or trials) to be successful in the treatment of D. viviparus. Charleston (1980) and Mason and Gladden

(1982), working in New Zealand, found that dosage rates used in the field ranged upwards from the levels recommended for cattle. Why these anthelmintics which are highly efficacious in cattle require to be used at higher dosage levels in deer has not been fully investigated but differences in the metabolism of these drugs by deer appear to be involved. Following treatment with the benzimidazoles, reinfection can result in patent infections approximately 3 weeks later. Ivermectin is currently widely used by deer farmers. Subcutaneous injection with the cattle formulation is the usual route of administration; dosage levels are commonly around 200 mcg Ivermectin/kg bodyweight. Following treatment with Ivermectin and continuing challenge from infected pasture, patent infections re-appear in approximately 4-5 weeks.

PROTOSTRONGYLID INFECTION

Two main protostrongylid parasites have been recorded in Scottish and European red deer: Elaphostrongylus cervi and Varestrongylus sagittatus.

Elaphostrongylus infection

E. cervi was discovered in Scotland in 1931 (Cameron) and is considered synonymous with E. panticola, E. rangiferi and Protostrongyloides cervi. It is one of the most common endoparasites of Central European red deer (Prosl and Kutzer, 1980). It also parasitizes various other subspecies C. elaphus (e.g. marals and wapiti), sika (C. nipon) roe deer (Capreolus capreolus), reindeer and caribou (Rangifer tarandus) and elk (Alces alces). Bovidae may be susceptible under certain conditions. Adult Elaphostrongylus live in the meninges over the brain and spinal cord or in the fascia of the muscles (the main sites being breast, shoulder, back and axilla). Prosl and Kutzer (1980) investigated the life cycle of E. cervi and showed that the eggs were laid into "the small blood vessels" and were carried to the lungs via the bloodstream. In the lungs the eggs grow, segment and develop into first stage larvae. The mean pre-patent period in experimental infection in New Zealand red deer is aproximately 112-114 days with a range of 107-125 days (Watson, 1983). Russian data, quoted by Prosl and Kutzer (1980), is similar (119-124) but in the caribou, Lankester (1977) found the interval to be 64-74 days. A wide range of molluscs has been found to

act as intermediate hosts for Elaphostrongylus in Austria, Eurasia and
North America (Table 2). In Scotland, larvae morphologically similar to

TABLE 2   Molluscan intermediate hosts of Elaphostrongylus cervi

---

Agiolimax agrestis   Lymnaea spp.
Arianta arbustorum   Mesodon thyroidus
Arion hortensis   Nesovitrea spp.
Arion silvaticus
Arion subfuscus
                                        Perforatella bicallosa
                                        Perpolita petronella
Bradybaena fruticum

                                        Radix ovata
Cepaea vindobonensis
Clausilia bidentata
Cochlicopa lubrica   Succinea altaica
Cochlicopa lubricella   Succinea granulosa
Cochlicopa pseudoniteus   Succinea pleifferi
Coretus corneus   Succinea putris

Deroceras laeve   Trichida hispida
Deroceras reticulatum   Triodopsis albolabris
Discus ruderatus

                                        Vitrina pellucida
Euconulus fulvus   Vitrina rugulosa
Euomphalia strigella

                                        Zenobiella aculeata
Galba spp.   Zenobeilla nordenskioldi
Galba palustris   Zonitoides nitidus
Galba truncatula

---

first stage E. cervi were found in the grey field slug (Agriolimax
reticulata) and the white-soled slug (Arion fasciatus). No doubt other
Scottish molluscs are involved.

Protostrongylid ova and larvae were discovered in 209 of 416 red
deer lungs examined in Scotland between 1977-84 (Munro, 1985).
Free-living and farmed deer were infected and no area of the country
appeared devoid of infection. Table 3 gives the rates of prostrongylid
infestation found during diagnostic histopathology (1977-84) and in
samples collected in the course of two surveys conducted in 1983 and
1984. Eggs and larvae in the lungs are usually restricted to animals

TABLE 3   Protostrongylid infection in Scottish red deer July 1977-
           August, 1984

Year	Farmed (%)		Free living (%)		Unknown
77	0/1				
78	5/9	(56)			
79	1/5	(20)	0/1		
80	2/6	(33)			0/1
81	6/11	(55)			
82	5/19	(26)	0/2		0/1
83 diagnostic	21/ 51	(41)	18/24	(75)	
83 (adult survey)	3/78	(4)	132/174	(76)	
84 diagnostic	3/14	(21)	2/2		
84 (winter survey)			11/17	(65)	
TOTALS	46/194	(24)	163/220	(74)	0/2

over 6 months of age and in such animals infection rates for the years
1977-84, in routine diagnostic samples from farmed deer ranged from
0-100% with a mean of 50.2%. Rates also varied greatly from farm to
farm.

Histopathology of E. cervi pulmonary lesions

The host response around the eggs is unpredictable. Some eggs
stimulate a slight mononuclear cell response within the alveolar septa
(Figure 6) and this is usually accompanied by collapse of one or more
adjacent alveoli. More intense infiltrations surround other ova and
cause greater thickening of the alveolar septa. In the more intense
infiltrations, neutrophils and eosinophils are irregularly present but
approximately 50% of infected animals have no significant eosinophil

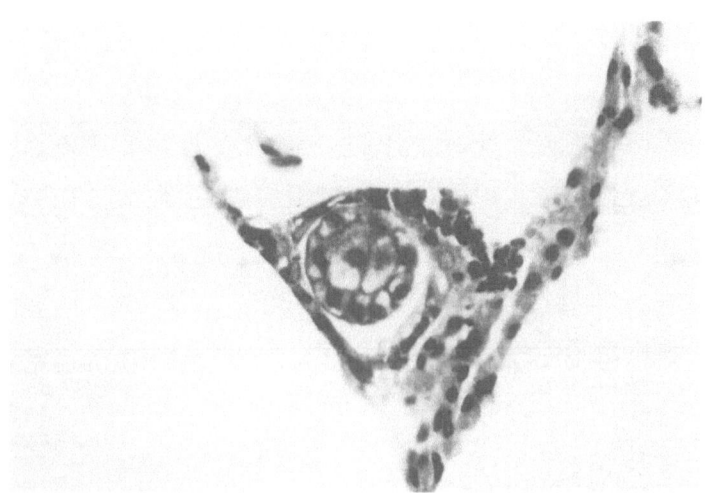

Fig. 6   Slight mononuclear cell resonse in the alveolar septa
around a developing protostrongylid egg.   HE x 380

infiltration of the lungs.   In those animals which show this response the
peri-bronchiolar and   interlobular   tissues   are   the   favourite   sites.
Lymphoid collections are present in the interlobular septa of some
animals but are generally of fairly modest size.

Twenty-five of the free-living deer examined by Munro (1985) showed
well-developed proliferative interstitial pneumonia asssociated with the
parasitism (Figure 7).   In each case,   large number of eggs and larvae
were present and they were surrounded by macrophages, round cells and
early fibrosis.   Alveolar   collapse,   thickening of alveolar septa
alveolar over-distension and emphysema accompanied and separated the
reactive foci. Eosinophils were irregularly present although they were
seldom a major part of the pneumonia. Mature larvae in the bronchiolar
lumina were   not   uncommon,   particularly   in   lungs with heavier
infestation,   but caused little irritation or cellular response.

Control of E. cervi infection

Three main factors influence the infection rate in farmed deer:
a)   Introduction of E. cervi via infected breeding stock.
b)   Availability of suitable intermediate hosts.

40

c) Anthelmintic treatments.

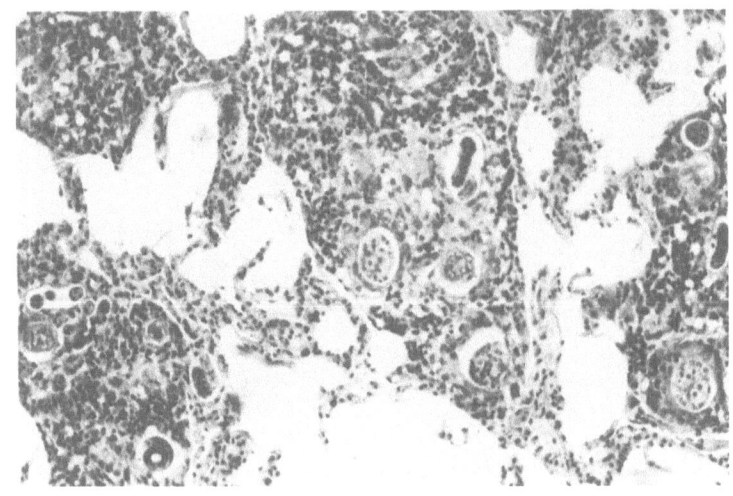

Fig. 7   Proliferative interstitial pneumonia associated with
protostrongylid eggs and larvae.   HE x 114.

Captured   free-living   deer constitute a major part of the   original
stock on many deer farms   and,   considering   the   level   of infection in
free-living populations,   infection must inevitably have been   introduced
on to these   farms.   However   clinical   signs   directly   attributable to
natural   E.   cervi   infestation have not been recognised   in   either   the
farmed or free-living populations. This in itself is not unexpected since
the   majority   of   infections   are   almost   certainly   sub-clinical.   No
information is available on   what   degree   of resolution of the pneumonic
lesions   occurs   following treatment.   The main economic impact   of   this
parasite may not, in fact, be a   result   of the pulmonary lesions but may
be   at   slaughter when greenish lesions which have developed   around   the
adult parasites in the muscle fascia have to be trimmed.

As   the   intermediate   hosts   are   ubiquitous,   control   relies   on
anthelmintic   treatment.   Three treatments on successive days   with   7.4
mg/kg/bodyweight per day or for   5   days   with 3 mg/kg/bodyweight per day
fenbendazole   will   kill the adult parasites.   Several weeks   may   elapse
after treatment before there is   a   significant   reduction in the rate of
faecal   larval   excretion.   Prosl   and   Kutzer   (1980)   made   the   very

reasonable suggestion that this delay indicates that, although the adult parasites are killed by the treatment, the eggs and first stage larvae in the lungs are unaffected and continue to develop - faecal larval excretion falling only as the lungs become progressively free of developing larvae. To ensure that faecal excretion is at a low level when the intermediate hosts become active in the spring, it is recommended that treatment should be given in December.

Varestrongylus infection

Varestrongylus sagittatus (synonyms: Bicaulus sagittatus and Caprcocaulus capreoli) frequently parasitises Europen red deer and is also the common protostrongylid of roe deer in Europe. Adult V. sagittatus live in the small bronchioles and give rise to nodules of verminous pneumonia. Blaxter and his colleagues (Blaxter et al., 1974) reported finding raised circular lesions beneath the pleura of red deer in Glensaugh. Older animals appeared to have heavier infestations than the younger stock and the eggs and larvae contained in these lesions were considered to be Bicaulus sagittatus. There have been no further descriptions of V. sagittatus in Scottish red deer. Three treatments each of 7.5 mg/kg/bodyweight fenbendazole are effective in removing V. sagittatus infections. Under farming conditions it is likely that anthelmintic programmes, using fenbendazole, instituted for the control of other endoparasites will also affect the protostrongylids.

REFERENCES

Blaxter, K.L., Kay, R.N.B., Sharman, G.A.M., Cunningham, J.M.M. and Hamilton, W.J. 1974. Farming the red deer. Department of Agriculture and Fisheries for Scotland, HMSO, Edinburgh. 70-76, 80.

Boev,S.N. et al. 1963. In "Helminthen der Huftiere von Kasakhstan". Alma-Ata. Cited by Enigk and Hildebrandt (1965).

Cameron, T.W.M. 1931. On two species of nematodes from Scottish red deer. Journal of Helminthology, 9, 213-6.

Cameron, T.W.M. 1932. Some notes on the parasitic worms of the Scottish red deer. Proceedings of the Royal Physical Society, 22, 91-97.

Charleston, W.A.G. 1980. Lungworm and lice of the red deer (Cervus elaphus) and the fallow deer (Dama dama): a review. New Zealand Veterinary Journal, 28, 150-152.

Corrigall, W., Coutts, A.G.P., Watt, C.F., Hunter, A.R. and Munro, R. 1982. Experimental infections with Dictyocaulus viviparus in vaccinated and unvaccinated red deer. Veterinary Record, 111, 179-184.

Corrigall, W., Easton, J.F. and Hamiton, W.J. 1980. Dictyocaulus infection in farmed red deer (Cervus elaphus). Veterinary Record, 106, 335-339.

Dunn,A.M. 1967. Endoparasites of deer. Deer, 1, 85-90.

Enigk, K. and Hildebrandt, J. 1965. Host specificity of Dictyocaulus species in ruminants. Veterinary Medical Review, 2, 80-97.

Erhardova-Kotrla, B. and Kotrly, A. 1973. The yearly cycle of some helminthoses of game animals from two ecosystematically different areas. Folia Parasitologica, 20, 41-48.

Lankester, M.W. and Northcott, T.H. 1979. Elaphostrongylus cervi Cameron 1931 (Nematoda: Metastrongyloidea) in caribou (Rangiferi tarandus caribou) of Newfoundland. Canadian Journal of Zoology, 57, 1384-1392.

McDiarmid, A. 1969. Diseases of red deer and their possible significance in deer farming projects, veterinary preventive medicine and public health. In "The Husbanding of Red Deer." Ed. Bannerman and Blaxter. Highlands and Islands Development Board and Rowett Research Institute. 50-54.

Mason, P.C. 1979. Lungworm in farmed deer. New Zealand Journal of Zoology, 6, 654.

Mason, P.C. and Gladden, N. 1982. Survey investigates drenching practice and internal parasitism on deer farms. Surveillance, 9, 2-3.

Munro, R. 1985. Histopathological observations on the lungs of Scottish red deer. D.V.M. Thesis. University of Glasgow.

Prosl, H. and Kutzer, E. 1980. Zur Biologie und bekampfung von Elaphostrongylus cervi. Zeitschrift fur Jagdwissenschaft, 26, 198-207.

Reenen, G. van 1982. Diseases. In "The Farming of Deer"(ed. Yerex) Agricultural Promotion Associates, Wellington, New Zealand. 119-134.

Volkholz, W. 1974. Untersuchungen uber der Lungenwurmbefall des Reh -, Rot - und Gameswildes. Inaugural dissertation, Tierarztliche Fakultat Munchen.

Waton, T.G. 1983. Some clinical and parasitological features of Elaphostrongylus cervi infections in Cervus elaphus. New Zealand Journal of Zoology, 10, 129.

Wilson, P.R. 1979. Diseases relevant to deer farming. University of Sydney Post Graduate Committee in Veterinary Science. Proceedings No. 49. (Deer Refresher Course) 105-117.

CRYPTOSPORIDIOSIS IN RED DEER

D.A. Blewett

Moredun Research Institute,
408 Gilmerton Road,
Edinburgh
EH17 7JH

ABSTRACT

Cryptosporidium is a small coccidian parasite which is not host specific and has so far proved insensitive to drug therapy. There is little information on cryptosporidiosis in deer, the infection has been reported in wild roe deer and has been shown to be implicated in illness and death in four outbreaks of enteric disease in farmed red deer. Data from an epidemiological survey of artificially reared red deer calves are presented together with comparable data from bovine calves. Although there were similarities in the prevalence and rate of spread of infection in both host species there were marked differences in the effects of infection. Out of 170 bovine calves monitored there were no losses attributable to cryptosporidiosis while there was good evidence that Cryptosporidium infection was the cause of death in 5 of the 30 red deer calves monitored, the possibility that parasite strain characteristics might influence the outcome of infection is examined and the problem of control are discussed.

INTRODUCTION

Cryptosporidium is a coccidian parasite. Although it has many characteristics in common with the Coccidia it also differs in several important respects, namely:

i] it has so far proved insensitive to the large number of anti-coccidal, anti-protozoal and antibiotic drugs which have been tested against it

ii] it is not host specific, infection is freely transmissible between different host species, including man. Cryptosporidiosis may be zoonotic.

iii] it has a very rapid life cycle, oocysts appear in the faeces only 4-5 days after infection

iv] it is an extremely small parasite with oocysts measuring just 5μm in diameter.

Infection with Cryptosporidium is a common cause of enteritis in young bovine calves and also occurs in lambs, kids and foals. The infection has been reported in wild roe deer but red deer are the only deer species in which clinical cryptosporidiosis has been shown to cause outbreaks of enteric disease. This paper describes an investigation of the epidemiology of cryptosporidiosis in artificially reared red deer calves

THE LIFE CYCLE OF CRYPTOSPORIDIUM.

Cryptosporidium has a simple, direct life cycle similar to those of parasites of the genus Eimeria. Infection is initiated by the ingestion of oocysts which excyst in the gut releasing sporozoites. The site of excystation is not known but in most host species infection establishes in the jejunum, ileum and caecum. In these sites the sporozoites invade the enterocytes of the villous epithelium where they take up a super- ficial position immediately below the cell membrane. Here the parasites multiply by at least two cycles of asexual division followed by differentiation into male and female gametocytes which then fuse to produce oocysts. A proportion of the oocysts (about 20%) are thin walled and excyst within the host thus maintaining the cycle of development, the majority of the oocysts produced are shed in the faeces providing a reservoir of infection in the environment.

The effects of Cryptosporidium infection in young ruminants are known best from studies of experimental infections in bovine calves. Using five day old calves infected with 10^7 purified Cryptosporidium oocysts there is a 4-5 day pre-patent period followed by an 8-12 day period of oocyst shedding. Total daily oocyst output is very large, between 10^9 and 10^{10} oocysts per 24 hours. Clinical signs generally coincide with the onset of oocyst shedding and consist of a moderate to severe scour, inappetance and dullness. These signs usually persist for 2-4 days although oocyst shedding continues for up to 10 days.

SURVEY OF CRYPTOSPORIDIOSIS IN RED DEER CALVES

Tzipori et al (1981) described clinical cryptosporidiosis in artificially reared red deer calves at the Macauley Land Use Research Institute's Research station at Glensaugh. Cryptosporidium was detected in deer calves on the farm in subsequent years and in 1986 a group of hand-reared calves was monitored over the first few weeks of life. Faecal samples were obtained from the calves on alternate days and screened for Cryptosporidium oocysts using the modified Ziehl-Neelson (MZN) technique. Calves which died during the survey were submitted for post mortem examination and samples of gut contents were screened for oocysts by the MZN technique and tissue samples from large and small bowel were processed and examined for histo-pathological evidence of cryptosporidiosis.

The prevalence and spread of infection

Of the 30 calves monitored, 4 died without showing signs of infection with Cryptosporidium. Of the remaining 26 calves, 24 became infected with Cryptosporidium, a prevalence of 92%. Initially, the infection spread slowly. Cryptosporidium oocysts were first detected in the faeces of two calves 10-12 days after the start of the survey. Since the prepatent period to oocyst shedding is 4-5 days and the first two positive calves had been housed for at least 8 days before they began shedding oocysts, it is almost certain that these initial infections were acquired in the calf house. Within 6-8 days of the first detection of Cryptosporidium infection the numbers of infected calves rose rapidly so that by 18-20 days after the start of the survey over 70% of the calves were shedding oocysts in their faeces.

Comparable surveys of bovine calves provide an interesting point reference. Week-old calves bought at market and transferred to calf rearing units show an immediate rapid rise in the prevalence of infection. In these cases up to 75% of calves may be shedding oocysts within 4-5 days of arrival on the farm, this is within the pre-patent period for Cryptosporidium oocyst shedding suggesting that the calves were infected before arrival. A survey of bovine calves reared on the farm in which they were born shown the same pattern of spread seen in red deer calves. There was a delay of 10-12 days before Cryptosporidium was detected then a rapid spread of infection until 100% of calves were infected.

The clinical effects of infection

Studies of experimental cryptosporidiosis in bovine calves have shown that the clinical signs of scour and inappetance occur during the first 2-4 days of oocyst shedding. Naturally infected red deer calves showed signs of scour and inappetance but these signs did not correlate closely with the onset of oocyst shedding. In three surveys of bovine calves (based on a total of 170 animals) the interval between the onset of scour and the onset of oocyst shedding was no more than 2 days in 50-75% of calves. In contrast, only 27% of red deer calves showed a similar temporal association between scour and oocyst shedding. Although there was no evidence of a clear association between oocyst shedding and scour in the red deer calves the mortality rate was far higher than that seen in bovine calves. Out of 170 bovine calves monitored in the three surveys, there were no deaths attributable to cryptosporidiosis whereas 5 out of 30 red deer calves died from the infection.

During the course of the survey 11 calves died, 4 of these were not infected with Cryptosporidium. Of the remaining 7 calves, all of which were infected with Cryptosporidium, 5 showed very similar properties, namely:-

i] they all began oocyst shedding at 6-8 days of age

ii] they all died 2-3 days after the onset of oocyst shedding

iii] they all showed scour and/or inappetance shortly before death

iv] they were all introduced into the calf house after the infection had become established.

Characterisation of the Cryptosporidium isolated from red deer

It is possible that the differences reported here between bovine and cervine cryptosporidiosis might be due to strain variation in Cryptosporidium isolates. This was tested by passaging Cryptosporidium isolated from a red deer calf in a bovine calf and comparing the pattern of oocyst shedding observed with that obtained using a bovine isolate. There was no evidence that the cervine isolate differed from the bovine isolate in either infectivity or pathogenicity. However, it was shown that the cervine isolate was more infective to new born mice than the bovine isolate. It is not yet possible to assess the significance of such small variations in the biological properties of different isolates of Cryptosporidium, but the existence of such variations must be borne in mind when comparing outbreaks of clinical cryptosporidiosis.

DISCUSSION

Little is known about cryptosporidiosis in deer. Tzipori et al (1981) described an outbreak of diarrhoea in 82 artificially reared red deer calves in which 56 calves were affected and 20 of them died. Cryptosporidium was detected in 80% of diarrhoeic calves compared to only 50% of normal calves. The calves became ill at between 3 and 5 weeks of age. Orr, Mackintosh and Suttie (1985) reported cryptosporidiosis in two groups of red deer calves in New Zealand. In the first group all 10 calves developed scour at a few weeks of age and died within a few days of the onset of illness. Cryptosporidium was identified in the only calf examined. In the second group 8 out of 10 calves developed scour at 9-15 days of age and 7 of them died within 2-3 days.   Cryptosporidium was identified in 5 of these 7 calves.

The epidemiology of clinical cervine cryptosporidiosis clearly needs further investigation.   The survey described in this paper indicated little correlation between oocyst shedding and diarrhoea, a mortality rate of 17% and suggested that calves infected when very young were at the greatest risk.   Tzipori et al (1981) found a stronger association between oocyst shedding and diarrhoea, a 25% mortality rate and severe clinical signs in animals between 3 and 5 weeks of age.   Orr et al (1985) reported an average mortality rate of 85% in animals of between 1 to 2 weeks and a few weeks of age.   The only consistent feature of the three surveys considered here is that they were all concerned with artificially reared red deer calves.   The level of stress induced by artificial rearing may be the major determinant of the outcome of Cryptosporidium infection.

Cryptosporidiosis is undoubtedly a serious problem in housed red deer calves, unfortunately the prospects for control are severely limited by the lack of an effective therapeutic agent.   Management and hygiene are the only options currently available.   Ideally calves should be kept in small isolated groups to limit the rate of spread of infection.   Such advice is likely to be impracticable but it is clearly inadvisable to introduce new born calves into an established group of 2-3 week old calves particularly if cases of diarrhoea have already occurred.   Hygiene is important in rducing the level of environmental contamination with oocysts. Cryptosporidium oocysts are very resistant but they are susceptiable to temperatures above $60^{\circ}C$, to strong ammonia, strong caustic soda and strong formalin (i.e. solutions of 5% or more).

REFERENCES

Tzipori, S., Angus, K.W., Campbell, I. and Sherwood, D. 1981.   Diarrhoea in young red deer associated with infection with Cryptosporidium. J. Inf. Dis. 144, 170-174.
Orr, M.B., Mackintosh, C.G. and Suttie, J.M. 1985.   Cryptosporidiosis in deer calves.   N.Z. Vet. J. 33, 151-152.

Conclusions from Session I

1. Papers on helminth parasitism in reindeer in the Netherlands, red and fallow deer in Denmark and red deer in Scotland highlighted the relative lack of information which is currently available on the production losses attributable to these infestations. It was clear from the papers and the discussions which followed that more work is required on the effects of management, housing, weather, winter feeding levels, concurrent infections, age and level of immunity on the establishment and clinical expression of parasitism in farmed deer. A multidisiplinary approach to these investigations is considered to be essential.

Reindeer in Denmark were performing poorly with few animals falling pregnant. In discussion it was pointed out that Reindeer at Whipsnade in the UK had performed in a similar unsatisfactory way until the feeding regime was altered to allow a much reduced winter food intake resulting in compensatory growth in spring and animals in much better condition.

2. For control of nematodes strategic dosing of infected groups of animals was recommended and maintenance of natural immunity by exposure to low levels of challenge was considered advantageous. Monitoring the faecal excretion of parasites is a valuable technique in formulating control procedures but further experience in the interpretation of faecal egg and larval counts is required.

3. Treatment of nematode infections with the benzimidazoles and ivermectin has been shown to be effective although the optimum dosage regimes still require to be determined.

4. Cryptosporidial infection has the potential to cause severe losses in deer calves. Experimental studies were considered to be urgently required since many fundamental questions regarding transmission of infection, immunity and control remain to be answered.

Chairman: Dr. F. Vigh Larsen

Co-Chairman: Dr. J. Fletcher

GENERAL AND ECONOMIC ASPECTS OF DEER FARMING

G. Reinken

Landwirtshaftskammer Rheinland, Endenicher Allee 60,
5300 Bonn 1, Germany

ABSTRACT

Grass and fallow land are extremely suitable for keeping fallow deer, but moorland and wet areas should not be used. According to our experience fallow deer are prone to liver fluke. They do not stay in water or wet areas for long. Woods should only be used in exceptional cases, for the animals nibble at the young trees and shrubs. On the other hand, neglected areas can be cleared by putting the fallow deer to graze on them.

Since these animals stay on the same plot of ground all year round the water supply must be secure. This can come from springs, brooks, or public water supplies. In any case, the supply of water and the water quality must be carefully checked before putting the fallow deer onto the land.

For economic and preservative reasons - not too much fencing - a minimum size of 4 hectares should be kept to. When keeping 12 to 15 female animals per hectare of land 2 to 2.5 hectares will be required, depending on the grass growth, and at a ratio of 1:30 male to female animals. The animals have enough space, the amount of fencing compared to the area is low, and the costs for the necessary buildings, for example, practice rooms, for fodder, water supply and silo, etc. are reasonable compared to the number of animals held. Since approximately 50% of the male calves, that is approximately 20 animals per year, are for meat production 1 to 1.5 hectares of land must be fenced in separately.

The economy of fallow deer farming is better than that of lamb and beef meat production.

INTRODUCTION

The permanent pressure of costs and the necessity of rationalisation gave rise in 1971 to the question as to whether the stock of useful animals had diminished and whether the alternative, cattle or sheep, was sufficient for making use of pastures and fallow land. We were of the opinion that the domestication of animals should not be brought to an end and began to look for another species of animal, in order to maintain the cultural landscape of the Federal Republic economic, with a minimum of expenses for the tax payer. The following criteria were of vital importance when making this study: long life, resistance to diseases, capability of survival in winter months (possibility of keeping the animals all year in the open, without a

53

stable), high productivity, easy calving, harmlessness, low fodder
requirements, good utilization of fodder, quick growth, good meat
formation, high quote of meat for cutting up for sale, excellent meat
quality.

As a result we put the following animals to a thorough test: fallow
deer, elk, roe, reindeer, red deer, wapiti and wild boar. Upon carefully
checking the results obtained from literature and various game preserves
at home and abroad it was found that under local conditions in the
Federal Republic the elk and the reindeer proved problematic with regard
to feed and keep. The roe was also eliminated, due to its very selective
way of taking up food. The wapiti was very impressive, due to its size,
but its fertility rate was very poor under our conditions. And this
animal can become very aggressive when fenced in.

The decision between red and fallow deer could first be made when
several factors had been considered. The decision was made in favour of
the fallow deer due to its capability to become accustomed to humans,
suitability for fodder with a high grass content, good utilization of
fodder, better quality meat and excellent capability of grazing on a
large area of ground. Low fences are sufficient for keeping these
animals together and large numbers can be kept on one area of ground.

FALLOW DEER FARMING

The investigation started in spring 1974 at the Research Station
Haus Riswick of the Rhineland Chamber of Agriculture on an area of 4.4
hectares with 28 does and 1 buck, in the autumn of 1974 on a commercial
farm at Marienheide, in the hilly region East near Cologne, on 5.5
hectares with 32 does and 1 buck.

The outer fence has a total height of 1.70 m; 1/50 m are for
netting. The distance between the horizontal wires increases upwards
from 7.5 to 12 cm. It is important that upper and lower wire of the
netting is extremely strong. The distance between the vertical wires
should be 30 cm. Each individual horizontal wire is reinforced at
intervals of approximately 100 m. The inner fences can be 1.40 to 1.50 m
high without bracing wire. The distance between posts is 10 to 20 m,
depening upon the land conditions; further apart on level ground and near
together on sloping ground.

A "treatment ring" is advisable for feeding the animals, for protecting them against wind and sunshine, and for observing, examining, catching and killing.

A so-called "funnel" can also be erected at low cost. The "treatment ring" and "funnel" should be in a central location, they are for looking after animals from the individual enclosures. It is advisable to take any further developments of enclosures into consideration when first building them.

We give the animals concentrated feed and, if necessary, water in the treatment ring so that they can accustom themselves to it at all times of the year. It also serves as protection against the wind. By concreting the base the droppings can be easily removed.

For feeding purposes we have a transportable, covered rack, in which straw, hay and concentrated fodder can be kept. It is important that the fodder remains dry in all weather conditions. Watering-places are also important for supplying the animals all year round. A suckling doe can drink up to 2 litres per day. The watering-places should also be accessible during the frost. When the ground water level is low they must be filled artificially. If water pipes are laid, they should be near the farmyard or house.

The development of the weight of the animals has been recorded by the Riswick Research Station since 1974. The weight of the full-grown does is approximately 50 kgs. Due to the growth of the foetus an increase in weight was recorded towards June, the main calving month. The male fawns weigh 4.6 kgs (4.3 - 5.0) and the female 4.5 kgs (4.2 - 4.9) upon birth. When suckled normally by the mother the female can weigh 20/28 kgs by the autumn. The male fawns weighed 26/31 kgs. During the winter months the weight stagnated, despite extra fodder, particularly with the male animals. Once the grass starts to grow in the spring an increase in weight can be observed. The year-old bucks weigh on average 52 to 53 kgs, approximately 3 kgs more than the average weight of their mothers. When 17 months old, the does obtained an average weight of 44 kgs. They were fully developed. From their birth in June until the beginning of October the female fawns put on a daily amount of weight of 152 grammes and the male fawns 207 g. When two years of age the male animals weighed between 60 and 64 kgs. The fully grown bucks can weigh up to 127 kgs.

The habits of these animals were observed and recorded during the experimental period. The fallow deer tend to be struck by panic and attempt to flee easier than red deer. For this reason they should never be driven, like cattle or sheep. The animals should be treated calmly and spoken to. This is important when transporting and letting them out of the crates or vehicles in which they are transported. They do not like a sudden change from light to dark.

At Marienheide the animals were provided with open pastures and also woods. At a later date part of a shed was opened up for the animals, but a large part stayed in the open even in snow up to 50 cm high. They made themselves a hollow in the snow. Only when it is extremely hot do the animals seek a shady spot. The does apparently anticipate weather conditions long beforehand. They were seen to seek a sheltered spot some time before a snow or rain storm started and always feed with their back to the wind. During the suckling period and the change-over to dry fodder (hay, straw) it is important that sufficient water is available. Up to 2 litres of water can be consumed per doe per day.

Damages due to decortication were minimal on the farm, and only in the first year. There seems, however, to be a connection between feeding and decortication. If the fodder has sufficient moisture, such as silage, it was not observed. The animals should not be given twigs and leaves for this only encourages them to decorticate.

They prefer to feed on weeds, such as ox-tongue, dandelion and camomile, and are partial to rose bushes, raspberries and blackberries. Even nettles and thistles are eaten from the flower downwards. They were observed to feed on areas of intensive growth first of all. Also, the animals feed on freshly manured areas without any problem. When there is only a thin layer of snow on the grass the animals scratch the snow away. Since they feed on grass in winter, whenever possible, attention should be paid to the fact that pastures are not completely grazed before the autumn commences.

They lose their winter coat by the beginning of May. By this time they have also lost their antlers. The fawns are born from June onwards, up to September. The year-old animals tried to suckle together with the new-born fawns. It is perhaps better to separate the year-old fawns from the herd before calving time.

The rutting period usually begins in mid October. Since our experiences we had losses of bucks with antlers as a consequence of fighting. Therefore the antlers should always be removed. This can be done by burning out, if possible when they are four months old. They can also damage the fencing. There is a danger that, due to the tameness, a human can be seen as a rival during the rut and attacked. And by removing the antlers there was no influence on the fertility or habits. During the winter months it was even quieter on the feeding area.

The does are driven for a long time during the rutting period until they are ready for copulation. The first rutting period lasts until mid November. From mid December until the beginning of January there is a second rutting period of those does which are not pregnant, and a third can follow in February. It is for this reason that fawns can be born up to the beginning of September.

The animals can become very tame towards humans when they are in constant contact with each other. The fawns, however, are relatively timid during the first year. By using a whistle, particularly together with some fodder, the animals can be attracted from quite a distance, independent of the persons voice. But a distance of 30 to 50 m is kept, most probably as protection against thieves, particularly in more remote enclosures. By enticing the animals with fodder during the summer months they can be attracted into the practice rooms where they can be observed more intensely, and, if necessary, weighed, measured or killed.

The does can be particuarly aggressive towards dogs. They even attack large dogs and try to kill them with their hooves. This was observed at different times in both enclosures. Fallow deer are, however, extremely sociable towards other animals, such as cattle, horses and sheep.

Several animals were able to escape on the farm due to a landslide and consequent damage of the fence. They tried, however, to get back into the enclosure. Similar observations were made on several farms. The aggressive nature towards dogs and also the attempts to return to the enclosure show respect of one's own quarters. The animals can be returned to their enclosure by enticing them with fodder.

The possibility and economy of keeping fallow deer in pastures and on fallow land is influenced by the meat quality, sale and price. Bucks were slaughtered at 3 different times of the year. The weight increased from 32 kgs in April to almost 53 kgs in July. The slaughter results varied between 55 and 58%. The weight of valuable parts, such as a leg, back and neck, also increased during that time. When slaughtering the bucks at the end of September instead of August there was no increase in profits and only a slight increase in weight.

An important fact is that 50% of the total weight are meaty and valuable parts. The comparable value of fattened calves amounts to 55-60%, of fattened lambs 45-50% and of pigs 45-50%. The percentage of 37-40% of legs is approximately 1-2% higher than that of fattened calves, almost 10% higher than that of lambs and pigs. And the profits made from slaughtering are on the same level as a good fattened calf, and higher than a fattened lamb (50%).

Comparisons of their economy have been made, but in order to make it exact the calculations were based on the fodder requirements for the full-grown and young animals of three species. Once larger amounts of the valuabel deer meat are available more exact comparisons can be made with lamb and beef. The results showed that fallow deer bring 2.227 DM gross margin per hectare, compared to cows with 2.396 DM and sheep with 483 DM. If a new stable building is necessary gross margin for fallow deer would remain, that of sheep be negative, and of cows be 1.546 DM/ha.

The costs for renewing the live stock, for water, electricity, medicine and veterinary services are at their lowest for fallow deer; however, the costs for fencing and gates and also the amount for varous special expenses are at their highest. A real comparison of all three species can first be made when the variable expenses for keeping sheep and cows in stables during the winter are also taken into account.

A valid and comparable calculation of their economy, of course, cannot take each individual case into account. Larger or smaller areas, the presence of buildings, good or poor grass growth, deviation in the meat prices, longer use of full-grown animals, less labour costs by putting up one's fences etc. play an important part towards the total profits. The individual location of the farms must be critically examined. And the suitability and capability of the manager is a must.

The number of fallow deer farms in the Federal Republic of Germany in 1987 was about 2000 with about 40,000 does. There are only a few farms with red deer.

REFERENCES

Reinken, G. 1987. Eamtierhaltung. Publisher E. Ulmer, Stuttgart,
2 Edition, pp. 320.

DEER FARMING IN DENMARK, WITH SPECIAL EMPHASIS ON THE MANAGEMENT AND HANDLING OF FALLOW DEER

Frank Vigh-Larsen

National Institute of Animal Science
Forsøgsanlæg Foulum
P.O. Box 39
DK8833   Ørum Sdrl.

ABSTRACT

    The development of the deer farming industry in Denmark is briefly described. In 1987 an estimated 10-12.000 does and 2.000 hinds are beeing farmed on 300-350 deer farms. A bill was adopted in 1987, setting the frames for the farming of deer in Denmark. Data on reproductive performance and growth rates on monitored fallow deer farms are presented, and it is concluded that results are unsatisfactory. Reasons for this are discussed, and basic rules, on which to base future management of fallow deer farms in Denmark, are drawn. Also, provosional basic rules for the handling of fallow deer are drawn. It is stressed, that deer on farms are to be fed and managed like other domestic animals, although they are not as yet "domesticated".

INTRODUCTION

    The farming of deer is a relatively new enterprise in Denmark. The first few farms were established around 1980, but during the last 3-4 years a rapid development has taken place. The Danish Deer Farmers Association has approximately 450 members, and it is estimated that there are 300-350 deer farms. Most of the deer on the farms have been imported, and the availability of breeding stock in other countries has been the major factor determining breed distribution on the farms. The most common species is the fallow deer, with an estimated 10-12.000 does on farms in 1987. Red deer are also beeing farmed, with an estimated 2.000 hinds on farms in 1987.

    The rapid development during the last 3-4 years has caused political interest in the matter, and a bill, setting the frames for the future farming of deer, was adopted by the danish parliament in april 1987. The major contents of the bill beeing a registration of all deer farms, a limitation of the species allowed to red and fallow deer, establishment of rules for fencing, slaughter and marking of deer on farms, and an official

recognition of deer on farms as beeing domestic animals, allowing farmers to benefit from normal tax rules applying for other
domestic animals.

First october 1985 the National Institute of Animal Science (NIAS) launched a research programme to establish guidelines
for the farming of deer in Denmark. The first two years the
programme has focused on identifying the major problems on danish deer farms. Some of these problems are now beeing further
investigated in intensive trials.

METHOD

A number of private deer farms have been included in the
programme. They have received regular visits by staff form
NIAS, who has supervised management procedures on the farms,
and collected all available data. Special emphasis has been
placed on collecting data to describe conditions on fallow deer
farms, and to introduce handling systems for fallow deer.

RESULTS AND DISCUSSION

This paper presents selected data from fallow deer farms,
and it should be noted that data are still limited.

Reproduction

In a system where meat is the major output, reproductive
performance is of utmost importance. Asher (1985) reports
high conception rates and high fawn mortality, leading to weaning percentages around 70% in New Zealand. Schick et al.
(1983) and Mulley (1984) reports weaning percentages around 80%
from West Germany and Australia respectively.

Data on reproductive performance on monitored farms in
1986 and 1987 are presented in table 1.

The calving percentage is only partly satisfactory at farm
C, while at the other farms it is too low. A number of factors
are considered to be responsible for this.

The nutritional level is generally low on the farms in the
autumn. Grass is often in short supply, the weather is cold and
rainy, and calves are not weaned until post-rut, if weaned at
all. The combined effect of these factors are expected to

TABLE 1.   Reproductive performance on monitored
fallow deer farms in 1986 and 1987.

Farm	Year	No of does	Calving %	Calf mortality, %	Weaning %
A	1986	88 (0)[1]	79.5	14.3	68.2
B	1986	40 (14)	72.5	10.3	65.0
C	1986	29 (7)	89.7	7.7	82.8
	1987	23 (0)	87.0	35.0	56.5
D	1986	54 (39)	83.3	26.7	61.1
	1987	38 (0)	73.7	46.4	39.5
E	1987	189 (0)	-	-	73.0

1) 2-year old does.

strongly affect the possibility of the doe conceiving early in
the breeding season, i.e. late October/early November.

Yearling does are often too small to have a realistic
chance of conceiving at 16-17 month of age. In Australia (English, 1984) and New Zealand (Asher, 1985) the treshold weight
for oestrus activity in yearling does is expected to be 28-30
kg. On farm B, the yearling does were weighed at 21 month of
age, and weighed on average only 32.6 kg. Only 7 (50%) of them
calved in 1986, and although data is limited it is supportive
of the above mentioned.

Finally, a lot of old, infertile does are present at
farms. Culling has not been carried out because of the present
value of breeding stock, and a general understocking of the
farms.

In 1987 farms B and C have had a very high calf mortality.
On these two farms a lot of calves have been borne at weights
around 2-3 kg, apparently 2-3 weeks prior to normal delivery.
These calves are too small to stand and suckle, and most frequently they die within 24 hours (see also English, 1984 and
Asher & Adam, 1985). This phenomenon has occured at a lot of
other farms as well, the reason beeing unknown. Furthermore the

summer of 1987 has been unusually wet and cold, and this has caused further deaths among newborn calves.

In 1986 a questionaire was carried out to get a broader view of the reproductive performance on danish deer farms in general. Data from fallow deer farms are presented in table 2.

Data in table 2 are supportive of the situation on the monitored farms, and generally it can be concluded that the reproductive performance on danish fallow deer farms is unsatisfactory.

TABLE 2. Reproductive performance on danish fallow deer farms, 1986, questionaire (Vigh-Larsen, 1987).

Herd size	No of herds	No of does	Calving %	Calf mortality, %	Weaning %
1-10	18	93 (20)[1]	86.0	8.8	78.5
11-20	16	237 (85)	83.1	12.7	72.6
21-	12	416 (117)	75.0	10.6	67.1
Total	46	746 (222)	79.0	11.0	70.2

1) 2-year old does.

Selective data from the monitored farms and the questionaire do suggest, that weaning percentages of 80 or above can be achieved if feeding and management is carefully adopted to fullfill the nutritional requirements of the doe, especially during the mating period (see also Mulley, 1984).

Growth

Data from West Germany (Schick et al., 1983), Australia (English, 1984) and New Zealand (Asher, 1985 and Gregson & Purchas, 1985) indicate that average weights at 15-16 month of age of at least 35 and 48 kg for does and bucks respectively can be achieved.

Table 3 presents selected data from weighings at the monitored farms, indicating a wide variation in weights between farms. Farm C generally presents good weaning weights and good growth rates until 16 month of age. Yearling does are well above expected puberty tresh hold, and bucks are well developed already at 10 month of age (see also table 4). During housing

TABLE 3. Selected results from weighings at monitored fallow deer farms, 1986 and 1987.

Farm	Weigh date	Sex	Age, month	No	Average kg	s.d.	Daily gain kg
C	2/2 '87(w)[1]	M	7	15	31.9	1.91	–
	14/5 '87(t)[2]	M	10	14	43.1	2.43	0.111
	7/10 '87	M	16	5	60.4	4.05	0.125
	2/2 '87(w)	F	7	9	27.3	2.83	–
	14/5 '87(t)	F	10	9	31.7	2.29	0.044
	7/10 '87	F	16	9	42.7	2.55	0.075
	7/10 '87(w)	M	3.5	6	25.8	1.60	0.213
	7/10 '87(w)	F	3.5	7	21.6	5.50	0.171
D	16/12 '86(w)	M	6	12	18.9	3.70	–
	23/4 '87	M	10	13	37.8	4.75	0.148
	16/12 '86(w)	F	6	18	17.1	2.91	–
	23/4 '87(t)	F	10	18	27.6	2.17	0.082
	12/10 '87	F	16	17	39.6	3.28	0.070
	12/10 '87(w)	M	3.5	8	18.3	3.01	0.150[3]
	12/10 '87(w)	F	3.5	7	13.0	3.27	0.096[3]
E	11/8 '87	M	14	23	42.5	4.13	–
	15/10 '87	F	16	49	31.9	2.46	–

1) Weaning and housing. 2) Turn-out. 3) Estimated.

calves have been fed concentrates ad lib., and feed conversion rate has been estimated to [11.7] SFU (Scandinavian Feed Units, 1 SFU = net energy content for lactation of 1 kg barley) per kg liveweight gain.

It should be noted that some results from the slaughter of deer on farms in 1986 and 1987 indicate that liveweights in

yearling fallow bucks above 55 kg may be associated with excessive fatness (see also Schick et al., 1983). This question will
receive further attention in future trials.

Data from farm D show that very poor weaning weights in
december can be transferred to good yearling weights by housing
and ad lib. feeding of concentrates until spring, followed by
good grazing. At this farm feed conversion rates for the winter
period was estimated to 6.7 SFU per kg liveweight gain. It is
important to note, that calves wintered indoors on a straw bedding need their hooves trimmed before turn-out.

Yearling weights from farm E are considered typical of
farms where calves are wintered outdoors on a low plane of
nutrition.

Results indicate that good weights at weaning and at 15-16
months of age can be achieved, but only if the animals are fed
to meet their requirements, especially in the winter period.

Slaughter

Deer at farms in Denmark are beeing slaughtered at the
farm. They are veterinary inspected before slaughter, and then
killed either with a rifle in the field, or with a captive bolt
in the yards. The animals are eviscerated in a specially designed and approved vehicle, inspected and stamped by a veterinarian, and transported to slaughter premises for further
processing. Data on slaughterings are presented in table 4.

TABLE 4. Data on slaughter of male fallow deer, kg.

Farm	Year	No	Age, month	Live Weight	Evisc. weight	% of lw.	Cold carcass	% of lw.
1)	1986	26	15	46.5	32.8	70.5	25.8	55.5
C	1987	4	10	45.0	34.5	76.7	27.1	60.1
E	1987	23	14	42.5	-	-	23.1	54.4
1)	1987	51	15-16	51.4	-	-	-	-

1) Data from non-monitored farms.

It is important to note that housed, well fed calves (farm
C) achieved considerably better slaughter weights at 10 month
of age, than calves out-wintered on a low plane of nutrition,
slaughtered at 14 month of age (farm E). Generally slaughter
weights are not considered to be optimal, and trials are now
under way to determine optimum feed level during winter, optimum slaughter weight and slaughter and meat quality as affected
by different feeding strategies and slaughter weights.

Economics

At present deer farmers are receiving approximately twice
the price pr. kg carcass weight, as compared to beef cattle and
sheep. At farms, where good reproductive performance and growth
rates are achieved, gross margins can be very attractive, even
if yearling does are sold for slaughter. An estimated 1.500
yearling bucks have been slaughtered in 1987, but in a few
years output will increase dramatically. This will demand marketing and processing of the product in order to maintain present prices. For this reason some 250 deer farmers have formed
a limited company, with the purpose of upgrading and processing
the carcasses, and establish outlets for the final products.

Handling

It is a widely held belief that fallow deer can not be
handled. However, this is not true. Fallow deer are very mobile
animals, and demand considerably better yards and stockmanship
than red deer when yarded and handled, but if conditions are
met they can be handled in big numbers without losses. At farm
E approximately 400 animals were handled (weighed, udder checked, parasite treated, and tagged with eartags and collars) by
four men at a rate of 40 animals per hour.

Guidelines for the handling of fallow deer can be summarised as follows:

- Fallow deer move readily through gates if used to them,
and can easily be moved from a paddock into a race, width at
entrance 4-5 m.

- The race should narrow down to a width of 2-3 m at the
entrance to the yards, and turn around at least 2 corners at

this point. Sides should be screened off (e.g. ply-wood, hes-
sian) for the last 30 m before yard entrance, and a screened-
off gate is a must where the screening-off starts.

- Yards should be at least semi-dark, and the last room
before the crush should be totally dark. This will keep the
animals from jumping when handled.

- Animals are led into other rooms, and finally into the
crush, by the use of light.

- Animals are handled singly, securely held in a crush.

- Antlered bucks (this includes spikers) should never be
yarded, and bucks should always be separated immediately after
yarding.

A slide series, showing the handling of fallow deer on
farm E, will follow after the presentation of this paper.

MANAGEMENT RECOMMENDATIONS

Reproductive performance as well as growth rates are generally unacceptably low on danish fallow deer farms. The reasons
discussed indicate, that this is mainly due to management problems. On the basis of observations made on monitored - as well
as other - farms, and the cited litterature, a few provisional
basic rules should therefore be drawn, on which to base future
management of fallow deer on farms in Denmark:

- In Denmark, fallow deer are probably farmed close to
their northernmost distribution. This calls for special attention on the farm.

- Fallow deer (and red deer) are not fabulous animals, and
will not survive and produce on farms without some degree of
management and supplementary feeding. Maintenance feed level of
a mature 50 kg doe has been calculated to 0.9 SFU pr. day. If
grass is in short supply over the winter, animals should be fed
to fullfill their requirements. If shelter is scarce, the feeding level should be raised.

- Handling is not at all impossible, and calves should be
weaned, tagged and treated against parasites, preferably pre-
rut in early October, or post-rut in early January. Male calves
can then be turned out to pasture and grazed on their own over

the summer. They can then be slaughtered at any time without interfering with calving, rut etc.

- If calves are not weaned and housed over the winter they should be selectively fed with concentrates, e.g. creep feeding. Calves need plenty of shelter all year round.

- If calves are not weaned pre rut, does should be flushed over the mating period (mid October-late December). Whole grain is a perfect supplementary feed - good for the animals (low risk of acidosis if slow habituation), easy to handle and cheap.

- Does should be culled if not delivering a calf in two consecutive years. Udder check in early October is the only means of determining in larger herds, but demands skilled operators.

- The farm should be subdivided into at least a summer and a winter paddock, as deer, independent of feeding level, will pick sprouts as they emerge in the spring, and hence reduce grass production considerably.

If these few basics are followed fallow deer can be succesfully farmed, with satisfactory reproduction and growth rates, and with a minimum of management and disease problems. It is important though to stress again, that success will only emerge if deer on farms are treated like other domestic animals. When they are behind fences, they must be carefully fed and looked after, as they are unable to seek food and shelter on their own due to the fences.

LITTERATURE

Asher, G.W. 1985. Reproduction of farmed fallow deer (Dama dama L.). In: Deer Branch Course No. 2. Proceedings of a Deer Course for Veterinarians. Ashburton, New Zealand. p. 107-125.
Asher, G.W. & J.L. Adam. 1985. Reproduction of Farmed Red and Fallow Deer in Northern New Zealand. In: Biology of Deer Production. P.F. Fennesy and K.R. Drew, eds. The Royal Society of New Zealand. Bulletin 22. p. 217-224.
English, A.W. 1984. Fallow Deer - Their Biology, behaviour and management in Australia. In: Deer Refresher Course for Veterinarians. Proceedings No. 72. University of Sydney. p. 285-304.

Mulley, R.C. 1984. The Reproductive Performance of Fallow Deer
in New South Wales. In: Deer Refresher Course for Veterinarians. Proceedings No. 72. p. 461-480.

Schick, R., H. Bogner, P. Matzke, W. Braun, G. Burgstaller & H.
Vollert. 1983. Untersuchungen zur Haltungstechnik und
Wirtschaftlichkeit der nutztierartigen Haltung von Dam-
wild im Vergleich Zur Koppelschafhaltung. Bayerisches
Landwirtschaftliches Jahrbuch. 60 (4). p. 396-455.

Vigh-Larsen, F. 1987. Reproduktionsmæssige forhold på danske
hjortefarme, 1986. Hjorteavleren. 3, 1. p. 5-6.

DIAGNOSTIC EXAMINATIONS OF AUTOPSY MATERIAL SUBMITTED
FROM FARMED DEER IN DENMARK

H. V. Krogh & A. Mikél Jensen

National Veterinary Laboratory
P.O. Box 373, DK-1503 Copenhagen V.
Denmark

ABSTRACT

During a two-year-period 151 farmed deer were examined for diagnostic purposes. The material included four different species: red deer 17, fallow deer 107, sika deer 11, roe deer 6. Seventy-two were adults (>16 months) and 79 calves and yearlings. The animals originated from 87 farms. The monthly number of examinations varied considerably, most deer were submitted in February, June and July.

Stress, incorrect management and feeding were considered the main causes of death in 74 deer or 49% of the total number. In these cases pathogenic bacteria and viruses were not demonstrated. Perinatal mortality was seen in 20 calves, their weight was often less than half the normal birthweight. Traumatic lesions caused by goring, fighting or clinging in the fence were found in 15 deer.

Specific bacterial infections of rather minor importance were: Pasteurellosis 5, Johne's disease 2, Yersiniosis 2, Salmonellosis 1, Actinomycosis 1. Pneumonia was demonstrated in 8 deer, and Pasteurella multocida, Corynebacterium pyogenes and Staphylococcus aureus were isolated from their lungs. Parasitic elements were demonstrated in 54 of 105 deer, and in 14 cases the parasites were regarded of primary importance, i.e., gastrointestinal strongyles 6, lungworms 5, liver flukes 3. Malignant catarrhal fever (MCF) was observed in 14 deer. MCF was the only viral disease found in the material.

INTRODUCTION

The farming of deer has in recent years been taken up as an alternative production in Denmark. This production is suitable for poor quality land and may increase in the future, if the farmers succeed in obtaining a reasonably economic profit. In part it depends on minimizing the losses caused by disease, mortality or decreased growth rate. The possibility of eradication or prevention of important diseases in farmed deer depends on knowledge of nature and cause of these disease problems.

Information concerning diseases in Danish farmed deer is very limited. The purpose of the present paper is to report the results of diagnostic examinations of autopsy material from farmed deer in Denmark.

MATERIALS AND METHODS

The material consisted of carcasses and autopsy samples from farms and deer parks sent to the laboratory for diagnostic purposes during a

two-year-period (November 1985 through October 1987). Both carcasses
and autopsy material were included.

The laboratory examinations performed were according to ordinary procedures used in other ruminants at this laboratory, i.e., autopsy of carcasses or pathological examinations of submitted organs followed by bacteriological, virological, parasitological and histopathological examinations.

RESULTS

A total of 151 deer were examined during the two-year-period, 67
carcasses and 84 organ samples. Four different species were identified
or stated in the material: red deer 17, fallow deer 107, sika deer 11,
roe deer 6, the remaining ten cases were organs from unstated species.
Seventy-two adults (>16 months) and 79 calves and yearlings were included
in the material. The animals originated from 87 different farms. The
submissions to the laboratory increased during the period, 61 deer were
examined from November 1985 to October 1986, and 90 deer from November 1986
to October 1987. The monthly number of samples varied considerably, most
animals were received in February, June and July, the lowest number in
May and November. Results of the laboratory examinations were divided
into six main groups.

1. Disorders related to management and feeding

In 74 or 49% of the total number stress, incorrect management or
feeding were considered the main causes of death (Table 1). Some of the
conditions with a protracted course were probably associated with factors
as cold, wet environments, poor nutrition or social stress. In such cases
emaciation and dehydration were seen together with enteritis or enteropa-
thy. Stress caused by capture, handling and transportation presented an
acute disorder and death within a few hours after onset of clinical signs.
The groupings are made according to the most important findings at the
post mortem examinations, the results of the microbiological and histpa-
thological examinations and the anamnestic informations. Pathogenic bacteria and viruses were not isolated from these cases. Perinatal mortality was seen in 20 cases. The calves were stillborn or a few days old.
The weight of the calves was only half the normal, and no milk was observed
in the abomasum. In some cases _Escherichia coli_ was isolated from mesen-

At histopathological examination sarcosporidia were found in the heart and skeletal musculature of six deer.

TABLE 4. Parasitological examinations of 105 deer

		Number of deer
Parasitic elements not demonstrated		51
Parasitic elements demonstrated in large numbers		14
Lungworms (Dictyocaulus)	5	
Liver flukes (Fasciola)	3	
Gastrointestinal strongyles	6	
Parasitic elements demonstrated in small numbers		40
Strongyles	37	
Nematodirus	5	
Strongyloides	2	
Trichuris	9	
Capillaria	10	
Coccidia	10	

5. Virological examinations

A number of animals was examined for different viruses (Table 5). BVD-, IBR-, BRS-, PI-3-, Rota- and Corona-virus were not detected in the material. At histopathological examination malignant catarrhal fever was recorded in 14 deer in two outbreaks, one in red deer (6 deer submitted) and one in sika (8 deer submitted).

TABLE 5. Number of deer examined for viruses

	Number of deer	
	examined	positive
BVD-virus	94	0
IBR-virus	12	0
BRS-virus	10	0
PI-3-virus	10	0
Rotavirus	27	0
Coronavirus	34	0

6. Miscellaneous

Lymphosarcoma with excessive enlargement of all lymph nodes was seen in a two-year-old red stag. Some cases of abortion were examined in relation to occurrence of pathogenic microorganisms. In one case only, Streptococcus zooepidemicus was isolated, in the remaining cases no bacteria, fungi or viruses were isolated.

DISCUSSION

The causes of death in farmed deer in Denmark have not been systematically examined before. The present material originated from 87 farms. Vigh-Larsen (1987) stated, that there are about 300 Danish deer farms, thus, 29% submitted material for examination during the two-year-period in question. The variation in the monthly distribution of the material reflected two of the main problems, i.e., perinatal mortality in the summer, and death caused by starvation and exposure to cold, wet climate in the winter. Different conditions including faulty management and feeding were demonstrated in 49% of the material. In 18 deer the pathological findings were in agreement with the winter death syndrome (Ross, 1986), and in 13 deer an acute haemorrhagic enteritis similar to stress enteritis shock syndrome (Buxton, 1986) was seen. From some of these cases Escherichia coli or Clostridium perfringes type A were isolated, probably with no pathogenic significance.

Acute metabolic acidosis caused by grain engorgement was observed comparatively often, nine cases from seven farms, a dominance of male deer did not occur.

Perinatal mortality was demonstrated in 20 calves, 17 were fawns. The syndrome occurred more frequently in 1987 (15 cases) than in 1986 (5 cases). The higher mortality in 1987 was probably due to a bad climate during the year with a cold winter, late spring and the extremely wet and cold summer months.

A single case of nutritional muscular dystrophy was found in a two-month-old fawn. Furthermore, a low activity of glutathione peroxidase was found in a blood sample from another deer. The farm problem was that only 30% af the hinds were pregnant and the calving season prolonged. In Britain, poor growth rates and wasting were identified as selenium deficiency, which responded to treatment with selenium (Fletcher, 1987). Further investigations of selenium-responsive unthriftiness in Danish deer are needed.

rium necrophorum were isolated from different pyogenic or necrotic infections related to skin, oral cavity, abdominal abscesses and uterus. From
eight cases of pneumonia Pasteurella multocida, Corynebacterium pyogenes
and Staphylococcus aureus were isolated alone or in different combinations.
From nine cases of intestinal infection the following bacteria were isolated: Mycobacterium paratuberculosis (2), Yersinia pseudotuberculosis (2),
Salmonella dublin (1) and Clostridium perfringens type C (4). Bacteraemia or
septicaemia with infection in all organs examined was associated with Pasteurella multocida (5) Streptococcus zooepidemicus (2) and Clostridium sordellii (2).

TABLE 3. Bacterial infections in 151 deer

	Number of isolates
Staphylococcus aureus	8
Streptococcus zooepidemicus	3
Actinomyces bovis	1
Corynebacterium pyogenes	14
Mycobacterium paratuberculosis	2
Yersinia pseudotuberculosis	2
Salmonella dublin	1
Pasteurella multocida	7
Pseudomonas aeruginosa	1
Clostridium perfringens type C	4
Clostridium sordellii	2
Bacteroides sp.	1
Fusobacterium necrophorum	7
Total	53

4. Parasitological examinations

More than half of the deer examined, 54 out of 105, harboured parasites (Table 4), but in 14 cases only, the results of the parasitological examination were considered of primary importance. Gastrointestinal strongyles
were demonstrated in six cases, lungworms in five and liver flukes in three
cases. In the remaining 40 cases the parasitological findings were regarded as of minor importance. Examination for cryptosporia was performed
with negative results in 21 calves, five to 30 days old.

teric lymph nodes, liver and spleen probably caused by agonal penetration and with no pathogenic significance.

Table 1. Disorders related to management and feeding

Post mortem examination	Number of deer
Emaciation, dehydration, enteritis	18
Abomasitis and enteritis	13
Abomasal and intestinal ulcerations	6
Ruminal and abomasal foreign bodies, geosediments	7
Acute carbohydrate engorgement	9
Nutritional muscular dystrophy	1
Perinatal mortality	20
Total	74

2. Traumatic lesions

Lesions probably caused by goring, fighting or clinging in the fence were observed in 15 animals (Table 2). Most often perforation of the abdominal wall was seen. The perforations were complicated by peritonitis or abscess formation between or in different organs. Bacteriological examination of these cases yielded Corynebacterium pyogenes, Staphylococcus aureus and Fusobacterium necrophorum. Internal haemorrhages with the abdominal cavity filled with blood were recorded in four cases and serious bites in one fawn. Two cases of multiple rib fractures and one case with fracture of the lower jawbone complicated with osteomyelitis were seen; from the latter Staphylococcus aureus and Pseudomonas aeruginosa were isolated.

TABLE 2. Traumatic lesions

Perforation of abdominal or thoracic wall	6
Haemorrhages and bites	5
Fractures (ribs and mandible)	3
Intestinal rupture	1
Total number of animals	15

3. Bacteriological examinations

Bacteria isolated from the material are listed in Table 3. Most strains of Staphylococcus aureus, Corynebacterium pyogenes and Fusobacte-

Some of the bacterial infections are noticeable. A septicaemic form of infection with Pasteurella multocida was found in five cases all in fallow deer and all in July and August. Four of the cases originated from one farm. At necropsy haemorrhages throughout the body, swollen spleen and liver as well as enlargement of the neck were seen, these findings are in agreement with Munro (1986), who considered stress in the form of overcrowding and bad weather as predisposing factors.

Two cases of infection with Clostridium sordellii, one in a red calf and one in a doe were recorded. This organism is a normal soil contaminant, which occasionally causes malignant oedema. Richards and Hunt (1982) isolated C. sordellii from a lamb with clostridial toxaemia, entry was via the portal system.

Infection with Salmonella dublin was found in one case only, despite S. dublin is a common pathogen in Danish cattle. In Britain, Salmonellosis apparently is rare in deer (McDiarmid, 1975; Fletcher, 1982).

Yersiniosis was found in two male calves, which were seven and nine months old, respectively. Haemorrhagic enteritis and enlarged mesenteric lymph nodes were observed. The disease appears to be precipitated by stress (Beatson and Hutton, 1981).

Clostridium perfringens type C causes struck in strongly fed sheep. In the present material the bacterium was demonstrated in four cases, two with concommittant metabolic acidosis.

Johne's disease was found in two imported red hinds from two farms (Jørgensen and Jørgensen, 1987). The symptoms were emaciation and chronic diarrhoea. Catarrhal enteritis without hyperplasia in the gut wall and pronounced hyperplastic and caseous lesions in the mesenteric lymph nodes were characteristic pathological findings different from the bovine type of Johne's disease. Mycobacterium paratuberculosis was not demonstrated in autopsy material from 16 adult deer from other farms.

Malignant catarrhal fever (MCF) was found in 14 deer from two outbreaks. The first started in October 1986 in a quarantine of 27 red deer, imported from Germany at the end of September. Sixteen died, and six were submitted for examination, MCF was recorded in all. The second outbreak was in a farm, where nine out of 11 sika deer died during a period of three months, in eight cases MCF was demonstrated. In the latter outbreak sheep and deer were kept together in the same fencing. Previous contact with sheep can almost invariably be established when MCF occurs (Reid et al., 1984).

Based on the present survey, it might be concluded, that apart from the outbreaks of MCF no viral diseases and few specific bacterial diseases were observed. The problems in Danish deer farming are mostly associated with inappropriate management. Therefore, a better health status may be obtained through improved management.

REFERENCES

Beatson, N.S. and Hutton, J.B. 1981. An outbreak of yersiniosis in farmed red deer. In "Proceedings of a deer seminar for veterinarians". (New Zealand Veterinary Association). pp. 136-139.

Buxton, D. 1986. Stress enteritis shock syndrome. In "Management and diseases of deer" (Ed. T.L. Alexander). (Veterinary Deer Society, London). pp. 96-97.

Fletcher, T.J. 1982. Management problems and disease in farmed deer. Vet. Rec., 111, 219-223.

Fletcher, J. 1987. Veterinary aspects of deer management 2: Disease. In Practice, 9, 94-97.

Jørgensen, J.B. and Jørgensen R.J., 1987. Paratuberkulose hos krondyr (Paratuberculosis in red deer). Dansk VetTidsskr. 70, 322-324.

McDiarmid, A. 1975. Some disorders of wild deer in the United Kingdom. Vet. Rec., 97, 6-9.

Munro, R. 1986. Bacterial infections. In "Management and diseases of deer" (Ed. T.L. Alexander). (Veterinary Deer Society, London). pp. 67-68.

Reid, H.W., Buxton, D., Berrie, E., Pow, I. and Finlayson, J. 1984. Malignant catarrhal fever. Vet. Rec., 114, 581-583.

Richards, S.M. and Hunt, B.W. 1982. Clostridium sordellii in lambs. Vet. Rec., 111, 22.

Ross, H.R. 1986. Winter death syndrome. In "Management and diseases of deer" (Ed. T.L. Alexander). (Veterinary Deer Society, London). pp. 93-96.

Vigh-Larsen, F. 1987. Hjorteproduktion (Deer production). (Landhusholdningsselskabet, Copenhagen). pp. 1-112.

SLAUGHTER OF DEER

Thomas L Alexander BVM&S MRCVS

Ministry of Agriculture, Fisheries and Food
St Mary's Manor, North Bar Within,
Beverley, North Humberside, HU17 8DN

ABSTRACT

As the concept of deer changed from game to that of a farmed species,
a demand grew for enhanced hygienic production of venison and for meat
inspection. There was also an awareness of the welfare implication in this
demand. Seasonality in deer is a major consideration. Some farmed deer
are slaughtered in the field. An increasing number of deer are being transported to an abattoir to facilitate hygienic practices and meat inspection.
The production of a satisfactory product is also dependent on both the
farmer and the haulier, in particular to ensure that the deer arrive at the
abattoir in a clean condition and with minimal stress.

THE BACKGROUND

Modern deer farming commenced in the United Kingdom in 1970. The
inaugural meeting of the British Deer Farmers Association was held in 1978
and four years later the British Deer Producers Society Ltd was formed to
market venison produced by members of the Society. In the same year the
first slaughter of deer in a commercial abattoir took place in Scotland and
subsequently in England in 1984. As the concept of deer changed from game
to that of a farmed species a demand grew for enhanced hygienic production
of venison and for meat inspection. This demand could more easily be met
in some established abattoirs, but there was an awareness of the welfare
implication. The fear threshold is raised in farmed deer accustomed to
being handled and flight is not an immediate response to the presence of
man. Nevertheless it is possible to cause distress when deer are placed in
an unfamiliar situation particularly if there is mishandling and where the
flight response is precluded.

FIELD SLAUGHTER

The most humane method of slaughter is to shoot the deer as it grazes
or during feeding provided a proven marksman is used with a suitable rifle
and ammunition. Where quiet deer present a stationary target at a range of
10 to 20 metres a frontal head shot is feasible. Semi-wild deer at up to
40 metres distance can be taken with a high neck shot to break the spinal
cord. During routine feeding it may be possible to shoot 10 or more deer
from a large group before the others become unduly disturbed. If too few
in a group are left they may become unsettled and try to escape. There is

also a risk of panic if the paddock used is too small.

But it is not possible to ensure public safety on every farm and stunning and bleeding may be problematic.

ABATTOIR SLAUGHTER

Subject to welfare safeguards it is considered acceptable for farmed deer to be slaughtered in abattoirs.

Seasonality

A major welfare factor in considering the abattoir slaughter of deer is seasonality. This has been previously recognised in part for deer shot in the wild resulting in the prohibition of the taking or killing of deer in Great Britain during certain periods. Such a prohibition no longer applies in Great Britain to farmed deer which are otherwise safeguarded. Hinds advanced in pregnancy and young calves have traditionally been safeguarded and should not be taken to an abattoir.

The annual antler cycle presents a particular problem because of the risk of damage to the growing antler and later of injury to other deer by the hard antlers. In contrast to the unacceptable removal of antlers in velvet, the hard, insensitive antlers can be removed, but this should be done in advance of transportation to allow any bruising to resolve. Deer in velvet should not be subjected to abattoir slaughter. Dis-budding of calves destined for slaughter rather than breeding can be achieved.

The seasonal change in behaviour of male deer presents a hazard not only to other deer but also to those handling them and male deer exhibiting rutting aggression should not be handled for abattoir slaughter.

Dominance

The maintenance of dominance and conversely respect is a major feature of social interaction in deer. Subdued behaviour or, where such option is open, flight is adopted unless the adversaries are well matched when fighting will occur. As respect for mankind diminishes, man becomes increasingly the subject of threats. Dominance must be maintained but with deference to the male in rut. To do so requires a knowledge of deer psychology. Most handlers wear protective equipment which not only gives personal protection but also enhances confidence.

Safeguarding carcase keeping quality

Those responsible for the deer up to the time of stunning and bleeding should ensure that it is presented in a clean condition. Access to wallows must be restricted prior to slaughter. By minimising stress the keeping quality of the carcase is further safeguarded. Stress causes glycogen depletion in the muscles. On death this glycogen would have been converted into lactic acid, creating an acid environment. Low pH does not favour bacterial growth. An average pH of 5.6 has been recorded for red deer calves and yearlings.

Shedding out

Calm, confident, competent handling is required during shedding-out of deer to ensure that the deer are subjected to minimum stress and that no bruising or other injury occurs. They should be kept in familiar groups and should not be in close confinement overnight since fighting can occur when they are left undisturbed.

Transportation

Suitable facilities are essential for loading when deer are to be transported. Where the facility does not enclose the rear of the vehicle there must be effective solid sides of an adequate height to prevent any attempt to escape - a minimum height of 2.4 m (approximately 8 ft) for red deer. Sticks and goads should not be used. To deter the deer from breaking back, personal shields can be used as a visual barrier. In addition to previous familiarisation, the deer should be segregated by species, age, sex and size and given sufficient space to lie down, get up easily to stand in a natural position and turn around. Divisions should be solid and of sufficient height to deter the deer from trying to jump over. A side hinged gate at the back of the vehicle is advantageous to prevent the deer from jumping off the ramp as it is being raised. To prevent bruising and other injury the interior of the vehicle must be free from projections and other hazards and the underfoot surfaces designed to prevent slipping. Bedding should be provided. Ventilation must be adequate and suitable. The driving must be sympathetic.

The Lairage

At the slaughterhouse suitable facilities must be available and the staff competent. Prior arrangements should be made for the arrival of the

deer to ensure slaughter without delay. The lairage and slaughtering
facilities should not be in use for other animals unless there is separation of the activities by solid walls.

As for loading an unloading facility must be provided. If the deer
are reluctant to unload they must be given time to become familiar with the
surroundings and because deer move more readily from dark to light the lairage should be lit, but not such that deer are faced with a direct light
source.

The passage and lairage pen walls and doors should be smooth, of sufficient height to prevent any escape (approximately 2.4 m (8 ft) for red
deer) and solid to a height of at least 1.2 m (4 ft). Floor surfaces must
be non-slip and adequately drained. Adequate ventilation must be provided
and fresh, clean drinking water available. The use of noisy fittings
should be avoided and where appropriate fittings should be baffled to
reduce noise.

Provision should be made for bullied, ill or injured animals to be
separated but such that they can still see other deer. Deer which are in
pain or distress or are suffering must be humanely slaughtered immediately.

Deer should be held in familiar groups and have sufficient space to
stand up, lie down and turn around easily. No more than twenty deer
should be kept in each pen. Sexes should be segregated where the deer are
sexually mature. After unloading into the pens, the deer must be given
sufficient time to settle before being handled further and if necessary
the lights should be dimmed. The deer should then be slaughtered without
delay and in general within 3 hours. No blood or other refuse from the
slaughterhall should be deposited in or allowed to flow into the lairage.

Stunning and bleeding

The stunning crate should confine the deer in such a way that the deer
is unable to injure itself. A diffused light source must be provided at
the front of the crate to which the deer will be attracted.

Prior to stunning deer should not be held in the approach passages.
Only one deer at a time must be placed in the stunning crate and only when
the way is clear for it to be stunned and then bled immediately. Stunning
equipment must be properly maintained and reserve equipment readily available. Stunning should be by captive bolt or by free bullet if captive bolt
stunning is unlikely to be effective. A frontal stunning position should

be used for captive bolt pistol stunning. Struggling is likely to be caused by manual restraint of the head. The slaughter of a batch must be arranged such that slaughter of the last deer is not delayed. Pithing after stunning and before bleeding is not essential. Subsequent to hoisting by a hind leg deer are bled in a manner similar to that used for cattle. Both forelegs are held to minimise the risk to the slaughterman.

Electrical stimulation

Electrical stimulation treatment to hasten rigor mortis development must only be carried out on animals which have been both stunned and bled.

Dressing

Dressing is generally similar to cattle. A particular problem with deer is contamination with loose hair and wallowing mud. It has been found that a large volume spray to flood wash the carcase is a means of removing hairs. Gross contamination must be trimmed off the carcase. The harvesting of the penis, one of a range of by-products, can result in contamination because of the need to keep the brush attached, but this can be enclosed in a clean polythene bag. The dressing percentage for rising two-year old red stags is 55%.

Meat Inspection

Meat inspection has followed existing protocols. The possible presence of tuberculosis has to be kept in mind. Blood splashing on the diaphragm and abdominal wall is sometimes observed, but the aetiology is obscure. Bruising provides a good indicator of the adequacy of the previous handling of the deer and a study of the age and site of lesions can result in the detection of the cause. Focal bruising can be caused by deer placing the forelegs on the backs of other deer or by butting with an antler stump.

Health monitoring can be carried out at the abattoir by the recording and feedback of pathological findings and the taking of liver samples for deficiency estimations.

Hanging

The carcases must be hung such that air circulates freely between them, drying the surface and cooling the carcase. If chilling is too rapid cold shortening of the muscles will occur and produce tough meat. Electrical stimulation of carcases immediately after slaughter hastens rigor mortis and assists in the production of tender meat.

CONCLUSION

The aspects of welfare and carcase production noted above must become part of deer slaughtering. Any suggestion of adverse welfare or unhygienic production could be damaging to the industry and spoilage of the meat is not in the economic interest of any producer.

Conclusions from Session II

The feasibility of farming fallow deer profitably was clearly demonstrated from studies that had been carried out in Germany. However slaughter was only carried out in the field in contrast to red deer which could readily be brought to a slaughterhouse where high standards of hygiene could more readily be achieved. Fallow deer could however be processed at an abattoir after being shot in the field when farms were situated close to a slaughterhouse and rapid transport could be arranged.

The economic viability of the industry depended on minimising losses and achieving maximum reproduction efficiency. On many deer farms in Denmark the health status of herds could be improved by attention to management in particular ensuring that appropriate feeding regimens were adopted. Monitoring of performance and health were essential if optimal results were to be achieved. Apart from poor performance associated with inadequate feeding a number of specific disease had however been recognised and it was considered essential to develop a knowledge of them if they were to be controlled.

<u>SESSION III</u>

Chairman: Dr. E. Thiry

Co-Chairman: Dr. D. Buxton

HEALTH ASPECTS OF DEER FARMING

E. Körner, J. Winkelmann

Tiergesundheitsamt der Landwirtschaftskammer Rheinland, Rodeweg 5 - 11,
5300 Bonn 3, FRG

ABSTRACT

A description of a more then 10 years experience as consulting veterinarian in deer farms is given. Results of autopsies, parasitological
investigation, clinical chemistry and important diseases of farm deer are
described briefly. Management measures and preventive veterinary procedures and their evolution with special respect to antler removal in deer
calf are outlined.

In all areas of animal production prophylactic health care is one major
prerequisite for high output. It also ensures the efficiency of production
and lays the basis for achieving high quality products. Therefore it is
important to have deep knowledge of diseases and the possibilities of
therapy. In addition skillful management is necessary. It is a fact that
in the only recent developed agricultural deer farming little experience
is available.

At the beginning the only source of information was literature on wild
game deer in nature, parcs or other types of artificial farming for hunting purposes. In addition experience was taken from observations of
comparable livestock farming e.g. sheep.

Our own experiences are based mainly on observations made during the
last 10 years through the work done as consulting veterinarian in test
farms on deer with the usage of natural and uncultivated pasture ground.
These tests were carried out in the area of the Landwirtschaftskammer
Rheinland an agricultural institution for a part of the federal state
of North Rhine Westfalia.

Since 1974 the animal health institut of the Landwirtschaftskammer Rheinland in Bonn has carried out postmortems of animals which came from
agricultural farms. A summary of the number of postmortems are given in
the following table.

TABLE 1   Postmortems of the animal health institute, Bonn, 1974-1985

Year	1974-76	1977-79	1980-82	1983-85	Total
Number of Animals	18	64	70	31	183

The split up into sexes of the postmortems and the age of the animals is shown in the following table.

TABLE 2

total	sex		age			
	male	female	-4 week	-1 year	young	old deer
183	78	105	45	95	12	31

The results of the investigations are summarized in the next table.

TABLE 3

Results from 183 deer autopsies

diseases of the alimentary-tract 67	enteritis acidosis and alcalosis of rumen
bacterial infections 98	E. coli, clostridia, necrobacillosis streptococci
external influences 26	accidents, foreign bodies in the rumen, heart failure under stress, hernia abdominalis, dentitis
parasites 30	pneumonia by endoparasites, enteritis by endoparasites, ectoparasites
metabolic failure/ poisoning 6	copper poisoning
others 57	premature birth, stress, liver-abscess, tongue-necrosis, etc.

From these results it becomes obvious, that in agricultural deer farming diseases of alimentary tract and bacterial infections play the major role in contrast to accidents and external violance.

In addition to the above mentioned investigations we also observed since 1974 the parasite situation of deer. The results are given in the next table.

TABLE 4
___

results of parasitological investigations 1974-85

	1974-76	1977-79	1980-82	1983-85	total	%
number of samples	286	522	257	104	1169	
Tricho-strongyloidea	219	316	161	52	748	64
Strongyloidosis	70	71	69	23	233	19,9
Cestodiasis	9	6	-	-	15	1,2
Distomiasis (liver fluke)	1	4	16	-	21	1,8
Pulmonary nematodiasis	29	81	55	7	172	14,7
Coccidiosis	139	191	157	41	528	45,2
negativ	33	107	49	39	228	19,5

1974-85	total number of samples	% negative	% positive
	1149	19,5	80,5

We also carried out hematological investigations in the years 1983-85.
This was done of blood samples from male deer calf about five month old.
The results are as follows:

TABLE 5
⎯⎯⎯⎯⎯
Hematological investigations of deers

	average	standard deviation
hemoglobin g/100 ml	17,87	2,13
PCV %	49,92	3,71
erythrocytes Mio/µl	13,71	1,44
leucocytes /µl	3620	
neutrophiles %	67	10
eosinophiles %	2	1,5
basophiles %	0,1	0,2
monocytes %	1,0	0,4
lymphocytes %	32	10

Clinical chemistry		
AST U/l	75	24
y-GT U/l	12	11
urea mg/100 ml	58	18
phosphor mmol/l	3,4	0,6
calcium mmol/l	2,3	0,2
copper µmol/l	129,3	30

In 1981 Matzke and collegues confirmed through their investigations
the findings we have had in our institute. In addition he pointed out
the necessity of hoof-corrections on the certain farming conditions.
Schellner and collegues of the animal health institute of southern
bavaria investigated 125 animals or organs from deer farms and
they found that about 40 % of the deer died of endoparasitosis and
20 % of bacterial infections. He also found listeriosis, septicemia from
E. coli,pasteurellosis and one case of salmonella infection.
In 1985 Davidson et al. found in 5 deers antibodies against epizootic
hemorrhagic disease virus and in two samples antibodies against
blue tongue virus.

In our own investigations colibacillosis was found several times
being mortal on deer calf in agricultural farms. This disease is
known to cause high losses on the young animals under farm
conditions. Different types of appearance of this disease is known.
E. coli exists in the alimentary tract of healthy animals and
becomes pathogen under certain environmental conditions like
humidity, mal nutrition, cold, unhygienic conditions. The time of
incubation is about several hours up to 3 days. The peracute form
of the disease which is known from calf is not yet discribed
in deer calf . But we observed acute forms with diarrhoea
of bright yellow colour, partly foamy and highly disturbed general
condition with rapid exsiccosis of animals. Untreated animals
would die in a few days.
Postmortem investigations show more or less highly inflamed mucosa
of the abomasus and the intestine. The intestinal lymphnodes are
swollen and there is hyperemia of the organs. Diagnosis is to be
ensured bacteriologically. The bacteria have to be tested in respect
to their antibiotic resistence to improve special therapy.

Prophylactic treatment which we know from other animals was
not effective in deer or it was not necessary. But it is highly
recommended to look after optimal hygiene, to ensure ballanced
nutrition and navel desinfection in connection with marking

the animals on the first day.

Special vaccines for deer calf are not available on the
market but it is possible to produce farm specific vaccine.

Necrobaccillosis is another problem in deer farming. We have
seen several death cases of 8 to 10 weeks old deer calf in
autumn 1978 due to this disease. These bacteria can be found in
the soil and in the alimentary tract of healthy animals and
they become infective under unaerogene conditions together with
certain environmental circumstances. Deep purulent changes can be
seen in affected organs. We have seen most of the times the mucosa
of the oral cavity, the tongue and throught as well as the foot
to be affected. We have seen death cases through this disease
on a farm which was before occupied by sheep which showed a lot
of cases of foot disease. The infection obviously takes place
during food uptake through small lesions in the oral cavity or
tongue. Necrotic changes are found in the mucosa of the throught
and in the liver. Because diagnosed in the late stage of disease
we were not able to cure part of the calf so they died within
a period of 3 to 4 weeks because the animals were not able to
take up food any more.

In recent years we saw more cases of this disease from different
farms. Mainly young animals were affected.

Therapy is difficult. We tried to claen and wash the mucosal
defect thorougly with appropriate desinfectives and local
sulfonamides and/or antibiotics together with systemic therapy.
Seperation of these infected animals from the healthy ones is
important. Vaccination is not known for the time being.
Pasture grounds which are already used by infected sheep should
be avoided.

Desinfection of stable, food and rest places and other
devices is absolutely necessary. In endangered herds strict
birth hygiene and control of the calf has to be applied.

The strength of resistance of the mucosa is mainly determined
through ß-carotin and vitamin A. Therefore prophylaxis
depends on a sufficient supply of vitamin A and ß-carotin
through additional diet.

In our investigations we also have seen some

Metabolic disorders and poisonings

Poisons or metabolic disorders of small ruminants as described
in literature do not have any influence on deer kept in farms.
They were not observed or described up to now. Malfunctions like
acidosis and alcalosis of the rumen was observed when there was
a rapid change of food bases. This leads automatically to the
recommendation to avoid drastic food bases changes within a short
period. For deer an adaptation time of about 14 days should be
the minimum.
Our investigations also showed that copper poisoning is an
important metabolic disorder in deer. As observed in sheep
high copper content of mineral food is responsible for poisoning.
In addition copper polluted normal food and grounds can also induce
poisoning. We have seen that after uptake of greater quantities
calfs were dead on birth or died after a few days.
The ones that survived for several days showed insufficient
development, lack of vitality, icterus and sometimes diarrhoea.
Liver tissue showed toxic degeneration at histological examination.
Copper content of the liver was at 1300 to 1450 mg/kg dried substance.
Normal values in the liver are between 11 and 57 mg/kg d.s. (average
25,5 mg).
A treatment of affected animals is not possible. So prophylaxis
is important that means feeding food or mineral diets with low
copper contents which we know from sheep already.

For the deer farmers we worked out some

Prophylactic measures

Health risks go hand in hand with increasing density
of animals in the farms. Also environment has a strong influence
on health. There is also considerable influence by economic
reasons like purchase and sale of animals the exchange and
the need of economical utilisation of pastures. Also
it is very important to take animal hygienic aspects into
consideration when planning and running a deer farm.

Very often the recovery of one diseased animal is a problem
in agricultural farming. Only in exceptions the treatment
of a single animal is justified. So the veterinarian can
be active only in the prophylactic health care. Skillful
consulting needs special knowledge and experience with respect
to food, breed hygiene, biology and farming of deer.
Hygiene is determined mainly by measures against bacterial
and parasite infections. Those infections and also damages
through food cause not only fatal losses but also reduces
the economic output. One also has to keep in mind that the
meat of diseased animals is also a risk for humans.

Prophylactic health care in deer farms is therefore not only
an ethical but also an economical driving force.

I would like therefore to present some prophylactic measures:

Fencing secures against uncontrolled invasion of other animals
and therefore against infections. In Germany rabies is a special
problem which necessitates measures against foxes and other rabies
carriers. Please keep in mind that the fences should not be constructed
in a way that panicing deer can be hurt themselves when running
against them.

Quarantine areas for newly acquired animals for observation and
investigation are very helpful to keep infections apart from the
herd. New animals should be kept in isolation for about 4 weeks.
During this period different examinations like hematology,
parasitology and necessary therapies can be worked out.
The quarantine area has to be secured in a way that neither
escape nor invasion nor contact with other animals will be possible.
There should be a seperate entrance with a desinfection mat in
front. The area should also be constructed in a way that allows
thorough cleaning and desinfection.

Permanent health control by investigations of all died animals
hand in hand with parasitological control investigations is
absolutely necessary. Animals which appear sick should be
seperated and investigated. In case results requires therapy
respective measures should be applied.

Restoration programs should not only include drug treatment
but also the interruption of the life cycle of parasite.
Changing the pasture grounds and other measures which reduces
humidity of the ground and fencing of potencial parasite biotopes
are also sometimes necessities to reduce parasite infections.
Anthelminthics for treatment of deer are available on the market.
They are usually given with food.

Special care is necessary with dead or visible diseased animals.
They can be carriers or distributors of virus infections e.g.
rabies or other epizootics. Therefore dead animals should
immediatly be brought to a veterinary institute to get a sure
diagnosis. Concerning the feeding of deer quantity, quality and
biological value is of great importance. The appearance of the
food should be supervised permanently.
Sometimes analysis of the food gives important hints and enables
the necessary addition of minerals,trace elements and vitamins.
The copper content has to be kept in mind.

The feed has to be done at fixed places which can be cleaned easily and a contamination with soil is avoided. A sufficient supply of good quality water is absolutely necessary.

The removal of the antler system of young male deer is necessary for prevention of injuries and accidents and therefore an important part of preventive measures.

Visitors have not to be allowed to enter the grounds either alone or with other animals.

The preventive measures should therefore include the following points:
1.: analysis of faeces every three month followed by treatment against endoparasites at least however twice a year
2.: sufficient amount of minerals, trace elements and vitamins during pregnancy
3.: new born male deer have to be marked the first day. The navel has to be controlled and desinfected.
4.: removal of all antler of male deer at the age of about 4 month
5.: treatment against ectoparasites if necessary

Removal of antler

is the last part of this presentation.
To avoid direct or indirect economical damage and protect persons who handle the animals the removal of all antler is absolutely necessary in agricultural deer farming. By removing the antler you can avoid injuries, skin diseases and abortions. Sometimes severe injuries especially during rut lead to death. Even female animals might be involved. The attacking male animals also endanger people being around. Therefore the removal is an essential part of managing a deer farm.
Immobilisation for the operation of the animals is done by a

veterinarian.

The hairless antler structures of the skin of young deer up to
3 month was tried to remove by cantery, burning or stamping out.
This procedure cannot be applied with deer calf because the
development of the antler does not come from the skin but from
the periosteum which therefore has to be removed. Cantery is not
efficient and leads to malformation of the antler.

Due to our experience an effective and permanent removal can only
be reached on young deer by burning. This method was developed by
Rosenberger and collegues for calf . The animal should have reached
the age of about 3 months. Satifying results have sometimes not be
reached with younger animals. We use for this performance an
electric stick provided with an excavation in the approximate size
of the antler structure with a round sharply edged border. This
instrument is available on the market. The burning was done by
turning the electric stick around his axes through the skin so that
it can be shaved off, lifted and removed as a stamped out round
disk from the underground. The periosteum of the forehead bone has
to be set free. We never observed any complications like bleedings
or others. If the procedure is quickly and perfectly done no
anaesthesia is necessary beside the sedation for immobilisation.
Two helpers fix the animal on the head and at the limbs and the
operator applies the heated stick exactly on the antler structures.
On 300 treated animals neither in the first nor in the second year
any antler formation was observed.

The removal of the antler of stags was done by wire saw. Here again
sedation is sufficient for the operation. Special anaesthesia is not
necessary as the operation does not cause any pain. We use the wire
saw ly LIESS which is known from obstetrics in large animal practise.

I hope that I have given you some information which you can use for
your work with deer farmers.

REFERENCES

Reinken, G. 1987. Damtierhaltung, 2, Auflage.
Matzke, P. über Gesundheitsprobleme bei Damwild in Mastgehegen,
     Bay. Landw. Jahrb, 59, 484-495, 1981.
Schellner, H.P. 1982. Krankheiten des Damwildes, Untersuchungsergebnisse
     von 1977-1982, Berl. Münch. Tierartzl. Wschr. 95, 293-294.

TUBERCULOSIS IN FARMED RED DEER

(CERVUS ELAPHUS)

Fiona A. Stuart

Ministry of Agriculture, Fisheries & Food
Central Veterinary Laboratory
Weybridge, Surrey, U.K.

ABSTRACT

Several outbreaks of generalized tuberculosis in captive and wild deer, caused by Mycobacterium avium-intracellulare, have been described in the U.K. indicating the species' relative susceptibility to mycobacterial infection. In 1985 the first isolation of M. bovis in farmed deer in the U.K. was made from a consignment of Red Deer (Cervus elaphus) imported from Eastern Europe. The single comparative intradermal tuberculin test showed a specificity of 61.3% and sensitivity of 80% when compared with subsequent bacteriological tests on tissues taken at post-mortem examination of the deer. The lesions of tuberculosis seen ranged from lung lesions, large abscesses on pleural surfaces and in mesenteric lymph nodes, to discrete focal lymph node lesions. The implications of the possible introduction of M. bovis infection into farmed deer herds in the U.K. are discussed.

INTRODUCTION

Mycobacterium bovis infection has not been previously reported in farmed deer in the United Kingdom, although it has been recorded in seven out of approximately 450 culled wild deer in the West Country (Reports, 1985 and 1986) and in 5 out of 130 free-living deer in Ireland (Dodd, 1984). M. bovis has occasionally been isolated from captive deer in zoos (Jones, Manton and Cavanagh, 1976). The increasing importance of deer farming and the concomitant import and export of live deer presents potential problems in the control of bovine-type tuberculosis in deer in the United Kingdom of the type that the New Zealand authorities have experienced (Beatson, 1985). This paper describes an outbreak of tuberculosis due to M. bovis in farmed Red Deer (Cervus elaphus) imported to the UK from an Eastern European country.

MATERIAL AND METHODS

Origin of Animals

The index case was a Red Deer stag imported from an Eastern European country 3 months previously. The stag died and generalized lesions of tuberculosis were found at <u>post mortem</u> examination from which <u>M. bovis</u> was isolated.

During subsequent epidemiological tracing two other batches of deer were found originating from the same Eastern European country during the previous 18 months. All the deer had been kept in quarantine for 3 weeks before being moved to several different deer farms. The post quarantine movements are summarised in Table 1.

TABLE 1   Epidemiological Tracing Summary of Imported Deer

Date of Release from Southampton Quarantine	Number of Deer	Species	
MAY 1984 (BATCH 2)	51	Red deer	→48 FARM A, E. SUSSEX → 3 FARM B, E. SUSSEX
APRIL 1985 (BATCH 3)	78	Red deer	→61 FARM A, E. SUSSEX →17 FARM C, GLOUCESTERSHIRE
JUNE 1985 (BATCH 1)	28	Red Deer	→ 8 FARM A, E. SUSSEX →20 FARM C, GLOUCESTERSHIRE (INDEX CASE)
	12	Fallow deer	→12 FARM A, E. SUSSEX

Batch 2 consisting of 51 Red Deer were moved to 2 separate deer farms. The majority were moved to Farm A where they were kept in a large park with 1200 other deer, mainly Sika (<u>Cervus nippon</u>) and Fallow (<u>Dama dama</u>). A random cull of four of the imported Red Deer in the park, 18 months post-import revealed one to have visible lesions of tuberculosis at autopsy. As a result of this finding a total of 36 imported Red Deer, believed to be the total survivors of those originally moved, and 378 in-contact deer from the park were culled. One Red Deer stag and one hind from Batch 2 on Farm B died 22 months and 30 months after import respectively. These together with culls from this farm were examined <u>post mortem</u>.   Farm B in East Sussex, also under restrictions, had culls examined.

Batch 3 consisted of 78 Red Deer, also from the same premises in Eastern Europe, which were moved after quarantine in two groups to Farm A in East Sussex and Farm C in Gloucestershire and kept in isolation prior to intended export to New Zealand 6 months later.

All the imported deer had been tuberculin tested with negative results before leaving their premises of origin.

Tuberculin Testing

A single comparative intradermal tuberculin test was carried out on 51 Red Deer from Batch 3 at Farm A, one month before the intended export to New Zealand.   The test was carried out using 0.1ml each of 1.0mg/ml bovine purified protein derivative (PPD) (Weybridge) and 0.5mg/ml avian PPD (Weybridge).

The test results were read at 72 hours and interpreted as described by Lesslie and Hebert (1975).   All other imported deer and contact animals unless previously culled, were similarly tested later as a result of the epidemiological tracing.

Post Mortem Examination

The number of deer examined <u>post mortem</u> from each batch is shown in Table 2.

From each of the deer examined, a pool of tonsils, retropharyngeal, bronchomediastinal and mesenteric lymph nodes were taken.   In addition tissues containing any suspicious lesions were taken separately.

Bacteriological Examination

Six slopes of modified 7H11 medium (Gallagher and Horwill, 1977) were inoculated with a suspension of ground up tissues from each animal, and incubated at 37°C for six weeks.   Colonies resembling <u>Mycobacterium</u> <u>bovis</u> were identified as described by Marks (1976).

Suspensions of tissues from the imported Red Deer were also subjected to biological tests as described by Pritchard and others (1986).   Tissues from the culled in-contact deer were pooled into groups of five for biological testing.

Histopathological Examination

Sections of tissues with lesions suspicious of tuberculosis were stained by Haematoxylin and eosin and Ziehl-Neelsen for histopathological examination.

RESULTS

Results are summarised in Table 2.

TABLE 2   Post mortem and cultural findings in three groups of imported Red Deer and their contacts.

Batch	Number of deer examined	Number with visible lesions of tuberculosis	Number of reactors to the tuberculin test	Number from which M. bovis was isolated
1	1 imported deer (index case)	1	NT	1
2	38 imported deer	5	NT	5
	378 in-contact park deer	0	NT	0
3	51 imported deer	13	28	20
	13 calves	0	NT	0
	NT = not tested			

At post mortem examination the index case from Batch 1 at Farm C showed widespread lymph node enlargement, extensive consolidation and caseation of the lungs, and miliary foci of tuberculosis in both kidneys. Mycobacterium bovis was isolated on culture. All other deer on this farm passed a subsequent tuberculin test.

At post mortem examination the Red Deer stag from Batch 2 at Farm B presented abscesses in the retropharyngeal and rumenal lymph nodes. Mycobacterium bovis was isolated on culture. The Red Deer hind from this farm had no visible lesions (NVL) and biological and cultural tests were negative for tubercle bacilli.

On Farm A the 51 deer from Batch 3 tuberculin tested for intended export to New Zealand presented 18 reactors and 10 inconclusive reactors using British standard interpretation (Lesslie and Hebert, 1975).   Under severe interpretation there were 27 reactors and one inconclusive reactor. A further 5 deer had small bovine reactions (1–2mm) and two deer were reactors to avian PPD. At post mortem examination of the 28 deer with bovine excess readings (27 reactors and 1 inconclusive reactor) 12 were found to have visible lesions of tuberculosis (VL) and a further 4 with NVL were found to be positive for M. bovis by biological tests. Of the 23 deer which passed the tuberculin test on severe interpretation, 4 were positive for M. bovis; one had VL, one had a haemorrhagic retropharyngeal lymph node, one had non-tuberculous peritonitis and cellulitis due to a fighting injury and one had NVL. Of the VL animals presented, the majority of lesions were abscess-like, up to 8 cms in diameter and were found in lymph nodes or adhering to the pleura of the thoracic wall or diaphragm. Lung lesions were observed in only two animals, both reactors, in which they were widespread and diffuse foci of necrosis. One further reactor had pus in the tonsillar region and enlarged bronchial lymph nodes and was positive for M. bovis by biological tests.

Thus a total of 20 deer (39.2%) from Batch 3 were found to be tuberculous. Sixteen of the tuberculous deer were reactors or inconclusive reactors under severe interpretation. The tuberculin test thus gives a relative sensitivity of 80% (16/20) and relative specificity of 61.3% (19/31). It has a predictive value in negative cases of 82.6%, and predictive value in positive cases of 57%.

Of the remaining 36 Red Deer in Batch 2 on Farm A, 4 were found to be tuberculous. They had abscesses affecting lymph nodes in the head, thorax and abdomen. The frequency with which lesions were found at various sites is shown in Table 3. None of the 378 culled in-contact animals were found to be tuberculous.

TABLE 3   The frequency with which visible lesions of tuberculosis were found at various sites in a group of nineteen imported Red Deer

Site of Lesions	Frequency	
Head, thorax and abdomen	5	(26.3%)
Thorax only	4	(21.1%)
Abdomen only	3	(15.8%)
Head only*	2	(10.5%)
Head and thorax	2	(10.5%)

Head and abdomen	1	(5.3%)
Thorax and abdomen	1	(5.3%)
Head and skin	1	(5.3%)
(* Includes a haemorrhagic lymph node lesion)		

Histological examination of the lesions showed a few to have the typical structure of a tuberculous granuloma with central necrosis and calcification surrounded by epithelioid cells, a few giant cells of Langhans and fibrosis. The majority of gross lesions were abscesses with no distinguishing features histologically, and in most cases only a few acid fast organisms were present.

During this investigation, 26 imported deer were found to be infected with M. bovis, 19 of which had visible lesions. Eleven of the tissue samples with lesions and all of the seven with no visible lesions were detected as tuberculous on biological testing only, not on direct culture.

One isolation of M. kansasii and one of M. avium-intracellulare group was made from the culled Fallow and Sika Deer in-contact with Batch 2. No visible lesions were present at autopsy.

DISCUSSION

This is the first report of bovine-type tuberculosis in farmed deer in the UK. It appears likely that the deer were already infected when imported, since the index case died with generalized lesions of tuberculosis within three months of arrival and disease was confirmed in all three imported consignments but not found in any other deer examined on Farms A, B or C. The remaining deer had been in the UK for between 6 and 18 months and some transmission of infection is likely to have occurred between them. The finding of infected animals with no visible lesions at autopsy is evidence of this. Cases of M. bovis in farmed deer have been reported in New Zealand (Beatson, 1985) and the USA (Stumpff, 1982), and there have been reports of bovine-type tuberculosis in captive deer in parks or zoos from several countries, including India (Basak, Chatterjee, Neogi and Samanta, 1975), the UK (Jones, Manton and Kavanagh, 1976) and the USA (Quinn and Towar, 1963; Scott and Goyings, 1965). M. bovis has also been reported in wild deer in Canada (Belli, 1962), Hawaii (Sawa, Thoen and Nagao, 1974), Ireland (Dodd, 1984), Switzerland (Bouvier, 1963), UK (Reports, 1984 and 1986), and the USA (Levine, 1934). There is no indication of M. bovis infection in indigenous farmed deer in Great Britain at present.

With the increasing popularity and intensification of deer farming, tuberculosis, particularly that caused by M. bovis, could, if introduced, become an important problem since deer seem to be relatively susceptible to infection (de Lisle and others, 1983).   The first case reported in farmed deer in New Zealand in 1978, was in a deer herd on land contiguous with an area with a known tuberculosis problem in both possums (Trichosurus vulpecula) and cattle (Beatson, 1985).   It is also probable that some deer farms in New Zealand were established using tuberculous feral deer (MAF, New Zealand, 1986).

The potential sources of M. bovis for farmed deer in the UK are tuberculous cattle, badgers and imported or native free-living deer. Imported infected deer may well present the highest risk, as native wild deer have been shown to have a very low prevalence of tuberculosis (1.6%) within well defined geographical areas of Wiltshire and Dorset (Reports, 1985 and 1986).   Stringent control measures are necessary to prevent importation of diseased animals, and MAFF now require, in addition to pre-import testing, a clear herd tuberculin test four months after importation.   Until this is completed deer are held in isolation at the receiving deer farm.

If tuberculosis becomes established in a deer herd it may possibly be detected by tuberculin tesing (e.g. for export purposes, or private testing for monitoring purposes), by the appearance of clinical cases in the late stages of the disease, or, more likely, at post mortem examination. Full investigation of all deaths in farmed deer must therefore be carried out assiduously and where deer are being slaughtered in a slaughterhouse and carcases are being inspected by a Local Authority inspector, maximum use of this information should be made.   All abscess-like lesions should be regarded as suspicious of tuberculosis and sent for bacteriological examination.   The medial retropharyngeal lymph nodes have been found to be the most common site of lesions in deer in New Zealand (Livingstone, 1980), but in the present study lesions were found to be more widespread (Table 3).

A voluntary eradication scheme for tuberculosis in deer is now in operation in New Zealand based on a tuberculin test and slaughter policy. There are problems with this approach because of the practical difficulties

of handling and testing deer (Beatson and Hutton, 1981), and also because
of a high incidence of false negative reactors in this test.   The latter
may be partly due to the relatively rapid spread of infection in deer (de
Lisle, Carter and Corrin, 1985).   A single cervical tuberculin test
using 0.1ml of 2mg/ml of bovine PPD is normally used in New Zealand and
any reaction is considered positive.   There is controversy about its
efficacy compared to the single comparative cervical test using bovine
PPD at a concentration of 1mg/ml and avian PPD at 0.5mg/ml as used in
Britain (Carter, Corrin and de Lisle, 1984; Beatson, Hutton and de Lisle,
1984).   As a result of this, since 1986, deer which are positive reactors
to the single cervical test and which come from a herd with a history of
non-specific sensitivity, are retested using the comparative test.

In the USA, Kollias and others (1982) found the comparative cervical
test using 3mg/ml bovine PPD to have 57% specificity and 84% sensitivity
in detection of bovine-type tuberculosis in deer.   In the present study
the standard British single comparative cervical tuberculin test, using
severe interpretation gave a specificity of 61.3% and sensitivity of 80%,
comparable with previous findings in cattle using standard interpretation
(Francis and others, 1978; Stuart, 1984).

In the UK, cases of deer with generalized lesions of tuberculosis
caused by organisms belonging to the M. avium-intracellulare group (some
of which are mycobactin-dependent) have been reported more frequently
than M. bovis in both wild and captive deer (Hopkinson and McDiarmid,
1964; Hime, Keymer, Boughton and Birn, 1971; Matthews, McDiarmid and
Collins, 1981).   Generalized disease due to M. avium group is very
unusual in cattle, and the detection of this organism in captive or
farmed deer herds is of interest and may be of significance.

The future situation in the UK may well mirror that which has occurred
in New Zealand where between 1970, when the first isolation of M. bovis
from wild deer was made, and 1983 a total of 504 deer was found to be
infected, 72% of which were farmed Red Deer (de Lisle and Havill, 1985).
Tuberculosis is now recognised as one of the most important diseases of
farmed deer in New Zealand, and strenuous efforts are being made to
control it.

Intensification of deer husbandry practices in the UK carries the
risk of allowing diseases such as tuberculosis to become established in
farmed deer (Matthews, McDiarmid and Collins, 1981).   This could present

a threat to the tuberculosis-free status of the cattle industry, the export trade in deer and in additon, it should not be forgotten that bovine-type tuberculosis is an important zoonosis. It is therefore essential that deer farmers and practitioners are aware of this potential risk, and investigate disease and death in deer conscientiously.

REFERENCES

Basak, D.K., Chatterjee, A., Neogi, M.K. and Samanta, D.P. (1975)
"Tuberculosis in captive deer". Indian Journal of Animal Health 14, 135-137.

Beatson, N.S. (1985)
"Tuberculosis in Red Deer in New Zealand". In 'Biology of Deer Production', Royal Society of New Zealand. Bulletin 22, pp 147-150.

Beatson, N.S. and Hutton, J.B. (1981)
"Tuberculosis in farmed deer in New Zealand". Deer Seminar for Veterinarians. New Zealand Veterinary Deer Association, pp 143-151.

Beatson, N.S., Hutton, J.B. and de Lisle, G.W. (1984)
"Tuberculosis - test and slaughter". Deer Branch of New Zealand Veterinary Association. Proceedings of a Deer Course for Veterinarians. pp 18-27.

Belli, L.B. (1962)
"Bovine tuberculosis in a White-tailed Deer (Odocoileus virginianus). Canadian Veterinary Journal 3, 356-358.

Bouvier, G. (1963)
"Transmission possible de la tuberculose et de la brucellose du gibier a l'homme et aux animaux domestiques et sauvages". OIE Bulletin 59, 433-436.

Carter, C.E., Corrin, K.C. and de Lisle, G.W. (1984)
"Evaluation of the tuberculin test in deer". Deer Branch of New Zealand Veterinary Association. Proceedings of a Deer Course for Veterinarians. pp 1-8.

de Lisle, G.W. and Havill, P.F. (1985)
"Mycobacteria isolated from deer in New Zealand from 1970-1983". New Zealand Veterinary Journal 33, 138-140.

de Lisle, G.W., Carter, C.E. and Corrin, K.C. (1985)
"Experimental Mycobacterium bovis infection in Red Deer". In 'Biology of Deer Production'. The Royal Society of New Zealand, Bulletin 22, pp 151-153.

de Lisle, G.W., Welch, P.J., Havill, P.F., Julian, A.F., Poole, W.S.H., Corrin, K.C. and Gladden, G.R. (1983)
"Experimental tuberculosis in red deer (Cervus elaphus)". New Zealand Veterinary Journal 31, 213-216.

Dodd, K. (1984)
"Tuberculosis in free-living deer". Veterinary Record 115, 592-593.

Francis, J., Seiler, R.J., Wilkie, I.W., O'Boyle, D., Lumsden, M,.J. and Frost, A.J. (1978)
"The sensitivity and specificity of various tuberculin tests using bovine PPD and other tuberculins". Veterinary Record 103, 420-425.

Gallagher, J. and Horwill, D.M. (1977)
"A selective oleic acid albumin agar medium for the cultivation of Mycobacterium bovis". Journal of Hygiene 79, 155-160.

Hime, J.M., Keymer, I.F., Boughton, E. and Birn, K.J. (1971)
    "Tuberculosis in a Red Deer (Cervus elaphus) due to an atypical
    mycobacterium". Veterinary Record 88, 616-617.
Hopkinson, F. and McDiarmid, A. (1964)
    "Tuberculosis in free-living red deer (Cervus elaphus) in Scotland".
    Veterinary Record 76, 1521-1522.
Jones, D.M., Manton, V.J.A. and Cavanagh, P. (1976)
    "Tuberculosis in a herd of Axis Deer (Axis axis) at Whipsnade Park".
    Veterinary Record 98, 525-526.
Kollias, G.V., Thoen, C.O. and Fowler, M.E. (1982)
    "Evaluation of comparative cervical tuberculin skin testing in cervids
    naturally exposed to mycobacteria". Journal of the American
    Veterinary Medical Association 181, 1257-1262.
Levine, P. (1934)
    "A report on tuberculosis in wild deer (Odocoileus virginianus)".
    Cornell Veterinarian 24, 264-266.
Lesslie, I.W. and Hebert, C.N. (1975)
    "Comparison of the specificity of human and bovine tuberculin PPD
    for testing cattle - 1. National Trial in Great Britain". Veterinary
    Record 96, 338-341.
Livingstone, P.G. (1980)
    "The evaluation of tuberculin tests in a tuberculous farmed Red Deer
    (Cervus elaphus) herd in New Zealand". Thesis for Master of
    Preventative Veterinary Medicine. University of California, Davis.
MAF, New Zealand (1986)
    Surveillance - Cattle TB issue. Animal Health Division 13, 12-15.
Marks, J. (1976)
    "A system for the examination of tubercle bacilli and other
    mycobacteria". Tubercle, 57, 207-225.
Matthews, P.R.J., McDiarmid, A. and Collins, P. (1981)
    "Mycobacterial infections in various species of deer in the United
    Kingdom". British Veterinary Journal 137, 60-66.
Pritchard, D.G., Stuart, F.A., Wilesmith, J.W., Cheeseman, C.L., Brewer,
    J.I., Bode, R. and Sayers, P.E. (1986)
    "Tuberculosis in East Sussex. III. Comparison of post mortem and
    clinical methods for the diagnosis of tuberculosis in badgers".
    Journal of Hygiene, Cambridge 97, 27-36.
Quinn, J.F. and Towar, D. (1963)
    "Tuberculosis problems at a deer park in Michigan". American
    Veterinary Medical Association Scientific Proceedings 110, 262-264.
Report (1985)
    "Bovine Tuberculosis in badgers". Ninth report by the Ministry of
    Agriculture, Fisheries and Food.
Report (1986)
    "Bovine tuberculosis in badgers". Tenth report by the Ministry of
    Agriculture, Fisheries and Food.
Sawa, T.R., Thoen, C.O. and Nagao, W.T. (1974)
    "Mycobacterium bovis infection in wild Axis Deer in Hawaii". Journal
    of American Veterinary Medical Association 165, 998-999.
Stuart, F.A. (1984)
    "Studies on the serological diagnosis of bovine tuberculosis". MSc
    Thesis. University of Surrey.
Stumpff, C.D. (1982)
    "Epidemiological study on an outbreak of tuberculosis in confined
    elk herds". Proceedings of the 86th Annual Meeting of the U.S.
    Animal Health Association. p 524-527.

Towar, D.R., Scott, R.M. and Goyings, L.S. (1965)
    "Tuberculosis in a captive deer herd". American Journal of Veterinary
    Research 26, 339-346.

PARATUBERCULOSIS

N.J.L. Gilmour

Moredun Research Institute, 408 Gilmerton Road,
Edinburgh, EH17 7JH, Scotland

ABSTRACT

The epidemiology, pathology, pathogenesis, diagnosis and immunology of paratuberculosis is reviewed as an aid to the understanding of the disease in deer. Paratuberculosis is a potential hazard in farmed deer as is shown by the description of an infected herd given in an appendix.

INTRODUCTION

Paratuberculosis (synonym Johne's disease) is a specific, infectious, chronic enteritis caused by Mycobacterium paratuberculosis (syn. johnei) and manifested by wasting and diarrhoea. It affects all ages other than the very young and is almost invariably fatal once clinical signs are present. Paratuberculosis affects domesticated cattle, sheep and goats but it has been reported in other ruminants, including buffaloes, camels, yaks and reindeer. It has occurred in parkland deer in England in 1907 and in European red deer on a game farm in Canada (Vance 1961). Exotic species of ruminants in zoological collections, including llamas, have been affected. Paratuberculosis is distributed worldwide especially in areas of intensive livestock husbandry and its spread has been hastened by the practice of importing pedigree animals to upgrade indigenous populations. Recent observations of farmed deer suggest that paratuberculosis is potentially an important disease. I have summarised below the salient features of paratuberculosis in cattle as an aid to the understanding of the disease in deer with the realisation that there will be differences in certain features because of the idiosyncratic response of deer to mycobacterial infections.

PREVALENCE

Here one must distinguish between infection and clinical disease since not all infected cattle develop clinical signs. In Britain in the 1950s M. paratuberculosis could be isolated from the mesenteric lymph nodes of between 6 and 17% of slaughtered cattle. Overall mortality was 0.45% of dairy cattle, reaching 5% on infected farms and the most

important cause of economic loss. In the Netherlands the life-span of
cattle on infected farms was one lactation less than on farms free from
the disease. The important difference between the infection rate and the
mortality rate will be considered under pathogenesis.

THE CLINICAL SYNDROME

The outstanding features are progressive wasting and diarrhoea. The
period between the appearance of the first signs and death are variable.
The disease may exceptionally run a relatively short course of weeks or
clinical signs may persist for over a year. Initially the diarrhoea is
intermittent lasting up to 3-4 days, and remission of wasting between
bouts of diarrhoea can occur. Eventually the diarrhoea is persistent with
increasing muscle wasting. Recovery is a very rare event. Initial
clinical signs are often noted a few weeks after parturition.
Non-specific signs such as loss of pigmentation from the hair and
submandibular oedema also occur. Appetite is usually not diminished and
fever is generally absent (see later).

MYCOBACTERIUM PARATUBERCULOSIS

The causal organism is a non-motile, Gram-positive, acid-fast rod
1-2 x 0.5 µm and it is distinguished from other mycobacteria by its
requirement of a factor provided by other mycobacteria (mycobactin) for
growth in laboratory media. Four- to 6-week old cultures at 37°C show
small, raised, dull-white rough colonies with irregular margins. A
bright yellow, very slow-growing variant has been described which causes a
distinct yellow lesion in affected sheep. Apart from the mycobactin
requirement the organism can be identified by the fact that it causes
paratuberculosis in experimentally-dosed ruminants and does not produce
tuberculosis in rabbits, chickens, and guinea pigs.

PATHOLOGY AND PATHOGENESIS

Gross lesions are present only in the intestines and the mesenteric
lymph nodes. The extent of the lesions does not always correlate with the
extent of microscopic lesions or the clinical severity of the disease but
gross lesions are seldom present in the absence of clinical signs of the
disease. The earliest lesions consist of plaques of thickening in the
terminal jejunum and ileum and can best be appreciated by holding the

opened intestine up to a light. The characteristic thickening and corrugation of the mucosa occurs in the terminal ileum, caecum and proximal colon. Enlargement of the mesenteric lymph nodes draining these organs is a more variable feature. Nodes may be enlarged 2 or 3-fold and on section the area of the cortex is greatly increased. Caseation and calcification are not features of bovine paratuberculosis, but are present in sheep and goats.

HISTOLOGY

The first stage in the development of the lesions occurs in the Peyer's patches of the intestines and the mesenteric lymph nodes. Small foci of epithelioid cells and a few Langhans-type giant cells appear in the germinal centres of the lymphoid tissue and prolonged searching may reveal that a small number of those cells contain one or two acid-fast bacteria. Also, the lamina propria of the small intestine is infiltrated by small numbers of lymphocytes, plasma cells and eosinophils. Progression of the disease is marked by an increase in the degree and extent of the pathognomonic, epithelial cell infiltration of the villi, the muscularis mucosa and the submucosa and associated distortion of villus structure. The epithelioid cells ultimately form a syncytium which is responsible for the grossly-apparent thickening of the intestinal walls. As the disease progresses the numbers of acid-fast organisms usually but not invariably increase until micro-colonies can be found both intracellularly and in advanced cases extracellularly. At that stage there is dissemination of M. paratuberculosis throughout the body even extending, in some pregnant animals to the foetus.

PATHOLOGICAL CHEMISTRY

There is a loss of plasma proteins across the intestinal wall and deficient absorption of amino acids. The resultant biochemical changes resemble those of human protein calorie malnutrition thus accounting for the weight loss which, in paratuberculosis, can amount to 30% of normal body weight.

PATHOGENESIS

The development of methods for counting the organisms and the administration of repeated, small oral doses of M. paratuberculosis to sheep and cattle at Moredun Research Institute led to a fuller understanding of the pathogenesis of the disease. Progressive infections were not established in organs other than the intestinal mucosa. In sheep as few as 10^3 organisms established transient gut wall infection and 10^9 organisms caused clinical disease. M. paratuberculosis multiplied in the small intestinal wall, and mesenteric lymph node infection was a result of drainage from that site. There was no evidence that the tonsils or retrophyarngeal lymph nodes were the primary portal of entry, and by three months after experimental infection there was bacteriological and histological evidence of infection and multiplication of the organisms. In calves similarly infected infection was also confined to the intestinal mucosa and mesenteric lymph nodes.

These and other findings support the hypothesis that in an infected environment all animals are exposed to varying degrees of infection and reinfection. Following initial infection the organism multiplies in the intestinal wall. Some animals recover from that infection, some remain infected sub-clinically for variable periods and others, after a period of sub-clinical infection, develop clinical disease and succumb. The development of lesions parallels that of the infection. Reinfection is apparently without effect but the size of the initial infection and the age at first exposure influence the final outcome - the earlier the exposure the more likely is the animal to develop clinical disease. Cattle first infected as adults very rarely develop the disease but may remain as sub-clinical carriers of infection. Differences in virulence between strains of M. paratuberculosis have not been reported so far.

Merkal and others (1970) suggested that the diarrhoea and the febrile episodes which they reported could be due to delayed-type hypersensitivity (DTH). In cattle with M. paratuberculosis infection and DTH the intravenous inoculation of johnin causes diarrhoea. Diarrhoea is abolished after the desensitisation which results from repeated doses of johnin and does not recur till hypersensitivity reappears. The cellular infiltrates into the intestinal walls also may be part of the DTH response. The extensive terminal lesions are therefore the final product of a complex interaction between causal organism and host response. It appears that the host's defence mechanisms can sometimes be advantageous and deleterious at others. The protective role of DTH was supported by studies of experimentally infected sheep. Those animals which quickly developed DTH had intestinal infections that were less severe than those of similarly treated sheep which did not develop DTH (Gilmour and others 1978).

EPIDEMIOLOGY

M. paratuberculosis is excreted in the faeces of infected animals, the number of organisms in the faeces reflecting the degree of intestinal wall infection. It follows therefore that the sub-clinical carrier excretes fewer organisms than the advanced clinical cases and it is the latter which are most important in the dissemination of infection. Infection is acquired by ingestion of M. paratuberculosis and as the organism remains viable for up to 1 year on pasture infection is readily spread. In advanced clinical cases M. paratuberculosis is present in the milk and approximately 1/3 of foetuses of advanced cases become infected in utero.

Calves and lambs in contact with clinical cases have therefore many opportunities to acquire infection and the earlier they become infected the more likely they are to develop clinical disease.

DIAGNOSIS

I do not propose to review here all the tests which have been used to detect the subclinically infected animal or to confirm the diagnosis of clinical cases. Instead I propose to examine the reasons why they have failed in their objective especially when used in attempts to control the infection on a herd basis. In the first place no test has been developed which can be relied on to detect subclinical carriers consistently or to predict whether clinical disease will develop. Animals that have apparently completely eliminated infectious organisms may also give positive reactions. Skin tests for DTH are intermittently positive from a few months after infection until the onset of clinical disease when animals tend to become anergic. The opposite progression is seen with tests for serum antibodies, which develop many months after initial infection and are positive intermittently. However, most clinically affected animals will have antibodies detectable, for instance, by the complement fixation test. Microscopic examination of faeces detects only about 1/3 of clinical cases and few carriers whereas systematic faecal culture on a number of occasions will detect most clinical cases as well as sub-clinical excretors. In addition, culture is the only specific test for <u>M</u>. <u>paratuberculosis</u> since both DTH and serum antibody tests give cross-reactions with other acid fast organisms and even with non acid-fasts e.g. <u>Corynebacterium</u> <u>renale</u> (Gilmour and Goudswaard 1972). Culture is however extremely cumbersome and the absence of more convenient tests to detect sub-clinical infection has contributed to the spread of paratuberculosis from country to country. There is still a real need for a simple specific test to detect subclinical infection, and preferably one which could be used to predict those animals which will develop clinical disease. Modern methods of antigen analysis might allow progress to be made in this direction.

VACCINATION

Vaccination with <u>M</u>. <u>paratuberculosis</u> in oil adjuvants is successful in preventing clinical disease but does not abolish the carrier state. A disadvantage of such vaccines is that they evoke both DTH and a serum antibody response which cannot be differentiated from the response to infection.

CONCLUSION

Paratuberculosis in deer is a potential problem for deer farmers. In the outbreak described below the disease affected the younger animals and, as with other species, was difficult to diagnose in its early stages and to control. At the present state of deer farming development, in particular with the dissemination of breeding stock, our inability to detect the subclinical carrier presents a constant hazard. Further research is necessary into methods of diagnosing and controlling paratuberculosis in deer.

REFERENCES AND FURTHER READING

Gilmour, N.J.L. 1985. In Handbook der bakteriellen Infektionen bei Tieren ed. H. Blobel and Th. Schliesser VEB Gustav Fisher Verlag Jena, pp. 281-313.
Gilmour, N.J.L. & Goudswaard, J. 1972. Corynebacterium renale as a cause of reactions to the complement fixation test for Johne's disease. Journal of Comparative Pathology, 82, 333-336. 333-336.
Vance, H.N. 1961. Johne's disease in a European Red deer. Canadian Veterinary Journal, 2, 305-307.

Merkal, R.S., Kopecky, K.E., Larsen, A.B. & Ness, R.D. 1970.
    Immunological mechanisms in bovine paratuberculosis.   American
    Journal of Veterinary Research, 31, 475-485.
Gilmour, N.J.L., Angus, K.W. & Mitchell, B. 1978.   Intestinal infection
    and host response to oral administration of Mycobacterium
    johnei in sheep.   Veterinary Microbiology. 2, 223-235.

Appendix to Chapter on Paratuberculosis
Experience on a Scottish Experimental Deer Farm
W.A.C. McKelvey

Macaulay Land Use Research Institute, Bush Estate, Edinburgh.

HISTORY

In April 1985, three yearling red deer were moved from the Glensaugh Research Station of the Macaulay Land Use Research Institute, to Edinburgh. Shortly after arrival in Edinburgh one of these animals started to lose weight rapidly. Despite anthelmintic treatment, weight loss continued and diarrhoea developed. The diarrhoea did not respond to antibiotic and astringent treatment and faecal smears revealed the presence of large numbers of acid-fast organisms. The animal was destroyed and culture of post-mortem samples of faeces and intestinal mucosa confirmed the presence of M. paratuberculosis.

During the remainder of 1985 five contemporary yearlings were slaughtered following the onset of weight loss and diarrhoea, and thirteen other yearlings were shown, following a survey of all animals of that age, to be excreting acid-fast organisms in their faeces.

In 1986 serological screening of all the Glensaugh breeding stock and yearlings (400 animals) was initiated, this work being carried out in collaboration with the Scottish Veterinary Investigation Service. During that year some 35 animals were identified as having serum antibodies to M. paratuberculosis. Eleven of these animals were subjected to extensive post-mortem examination and 8 showed histological evidence of mycobacterial infection of the gut and mesenteric lymph nodes. Five other yearling animals showing clinical evidence of weight loss and diarrhoea have been slaughtered and all have been found to be infected with acid-fast organisms. Furthermore, all the yearling stags marketed through the local slaughter house over this period have been examined for histological evidence of mesenteric lymph node infection; 24/59 (41%) of these animals, all of which were clinically healthy, had acid-fast organisms in the mesenteric lymph nodes.

POST-MORTEM FINDINGS

The carcases were underweight, sometimes there was faecal staining

121

of the hind quarters and generally the coat was dry, easily plucked out and there was often patchy alopoecia. On opening the abdomen there was occasionally an excess of clear peritoneal fluid and in most cases the gut, on a casual examination, appeared quite normal. However, once opened, the gut presented irregularly thickened and reddened areas of mucosa predominantly in the lower small intestine, ileocaecal valve, caecum and occasionally in the proximal colon. The mucosa was not ulcerated. Most mesenteric lymph nodes were moderately to grossly enlarged, firm and yellowish on sectioning but seldom necrotic. No evidence of caseation or calcification was found in any of the nodes examined.

Histologically the mucosa and submucosa of the gut were heavily infiltrated with large macrophages laden with acid-fast organisms. The mesenteric lymph node sections showed evidence of hyperplasia with acid-fast organisms packed into macrophages and free in the stroma of the gland. One animal had gross and histological evidence of bronchial lymph node involvement.

This appendix is reproduced by kind permission of Dr. H.W. Reid, Editor of the Publication of the Veterinary Deer Society, 1987, Volume 2 (6) pp. 24-28.

SEROLOGICAL SURVEY IN FREE-LIVING RED DEER (<u>Cervus</u> <u>elaphus</u>) IN FRANCE

J. Barrat*, Y. Gérard*, A. Schwers**, E. Thiry**, J. Dubuisson**,
J. Blancou*

* Ministère de l'Agriculture, Direction Générale de l'Alimentation,
Centre National d'Etudes sur la Rage et la Pathologie des
Animaux Sauvages,
B.P. 9 - 54220 Malzéville, France
** Department of Virology and Centre de Médecine du Gibier,
Faculty of Veterinary Medicine, University of Liège, B-1070 Brussels,
Belgium.

ABSTRACT
Between March 1982 and June 1985, 89 sera were collected from free-living Red deer (<u>Cervus</u> <u>elaphus</u>) captured in the "Haute-Marne" region in France. These samples were principally used for a serological survey of the geographical distribution of antibodies directed against different antigens, in this species, in France. 46 antigens were tested. Although positive reactions were observed, none of them seemed to be related to clinical diseases in this population.

INTRODUCTION
The Centre National d'Etudes sur la Rage et la Pathologie des Animaux Sauvages, collaborating with other laboratories in France or abroad, carries out different studies on wildlife (Blancou, 1984). <u>First</u> : epidemiological survey in order to assess the situation of wildlife in general in relation with different infectious diseases. <u>Second</u> : biological analysis to estimate and follow the health status (e.g. : indices of condition, blood disorders : biochemistry and hematology) of these wild populations.
The results obtained during such epidemiological survey of a free-living Red deer (<u>Cervus</u> <u>elaphus</u>) population in France are reported in this paper.

MATERIAL AND METHODS
Population studied
The Red deer population studied here lives in the Arc en Barrois region (département de Haute-Marne) near Chaumont in the north-east part of France (see the map). The study area covers 135 km2 including 110 km2 of woods. Three wild ungulates are found in this region : Red deer, Roe deer (<u>Capreolus</u> <u>capreolus</u>) and Wild boar (<u>Sus</u> <u>scrofa</u>).

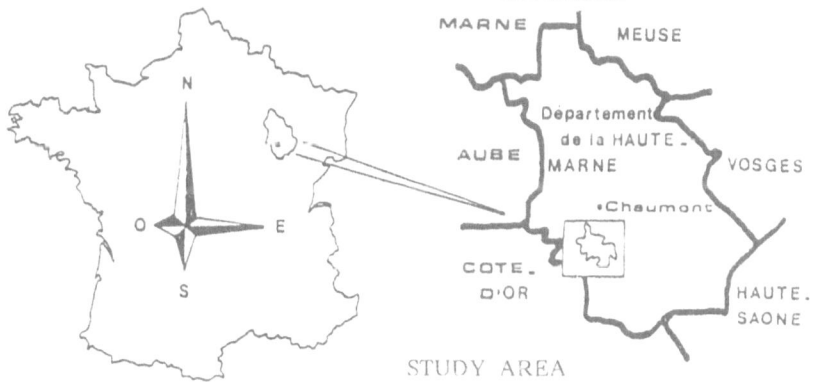

STUDY AREA

In 1971 the Red deer population was estimated to reach 800 animals, 1200 in 1976-77. Because of damages caused to woods and agriculture by the Red deer, the number of animals shot during the hunting period was increased since that moment. The present population is thus now supposed to range from 600 to 800 animals.

The average weight of adult females is 70 kg and the average weight of males is 145 kg (maximum 200 kg). Eighty per cent of young females and ninety per cent of adult are pregnant every year, and the sex ratio was estimated to be 1 male/1 female in 1982.

Capture and sampling methods

In a first step the whole study area (2,5 to 3 km2) is surrounded with ropes and flags. Then this study area is divided with two crossing lines of 10 to 15, 50 m long capture nets, against which the Red deers are beaten.

The contention of animals caught in the nets is only physical, nor chemical anaesthesia nor tranquillization are used during this study.

After that different measurements and samplings are carried out, the animals are released out from the sector.

Blood samples are taken by punction of the jugular vein. The volume of blood taken was 50 to 75 ml. An aliquot of serum is kept in our serum collection and frozen, that allowed different retrospective studies. Nasal swabs were collected for research of influenza virus and Probang tests were performed for foot and mouth wirus isolation. Feces were also collected for parasitology.

A rough clinical examination was eventually performed before the animals were weighed and released.

Serological analysis

As far as possible every serum sample was screened against the 46 antigens listed in tables 1 and 2.

Diseases possibly common to Red deer and domestic animals (foot and mouth disease, porcine influenza, IBR, bovine leucosis, mucosal disease complex, paratuberculosis, etc.) and/or Man (brucellosis, chlamydiosis, Q fever, influenza, leptospirosis, rabies) were first investigated. Two retrospective studies were also performed from the "serotheque". The first one screened antibodies against 14 arboviruses and the second one against 6 Herpes viruses (BHV1, 2, 4 and 6, HVC 1 and Reindeer Herpes virus). The results of this last work will be presented by Dr. Thiry during this seminar.

The serological techniques used during these studies are classical. They were previously listed, in technical references in Blancou, 1983. The other ones, concerning arboviruses may be found in Le Lay-Rogues and coll., 1987.

The thresholds considered as positive in this study are the ones classically admitted for domestic animals.

TABLE 1   Negative results obtained during the serological survey
----

Diseases or pathogens	Number of analysis
Arumowot	70
Bhanja	70
BHV 2	79
BHV 4	79
BHV 6	80
Chikungunyia	70
Dengue 2	70
Porcine Influenza	40
Texas (human) influenza	40
IBR (BHV1)	89
Leptospirosis	86
Bovine leucosis	21
Mucosal disease complex	23
P13	89
Rabies	63
Sindbis	70
Tahyna	70
Tick borne encephalitis	70
Tribec	70
Uukuniemi	70
Wesselsbron	70
West-Nile	70

RESULTS

The negative reactions are listed in table 1, and the positive in table 2.

TABLE 2   Positive results obtained during the serological survey
----

Antigens	Positive/Number of analysis
Adenovirus	33/89
Brest AN/219	1/70
Brucellosis	2/54
Chlamydiosis	5/54
Dermatophilosis	14/77
Eyach	1/70
Foot and mouth disease (O,A,C)	1/88
Q Fever	1/54
Bangkok (human) influenza	3/40
URSS (human) influenza	5/40
Paratuberculosis	9/52
Pasteurella multocida A	63/79
Pasteurella multocida D	9/77
Red deer HV (HVC1)	1/80
Reindeer HV	1/80
Rotavirus	54/71
Sandfly fever Sicile	1/70

Most of the reactions were observed against three pathogens : <u>Adenovirus,</u> <u>Rotavirus</u> and <u>Pasteurella</u> <u>multocida</u> A. Such reactions are often observed in other domestic or wild mammals. <u>Pasteurella</u> <u>multocida</u> A, for instance, is known to be a classical host of the respiratory system in birds and mammals without clinical signs.

The position of <u>P. multocida</u> D is quite different because this last one is often isolated in case of respiratory disease. Nine out the seventy-seven tests gave a positive result (i.e. 11,7%). This fact can be compared with the results of parasitological coproscopy (47 out of 56 examinations showed the presence of lungworms larvae). So, although, no abnormal clinical sign was detected during the examination of animals, a respiratory pathology involving viruses, bacteria and parasites might have been initiated in these subjects.

Two seroconversions against strains of <u>human</u> <u>influenza,</u> Bangkok (3/40) and URSS (5/40) were found. These positive reactions were observed both with radial hemolysis and hemagglutination inhibition reactions. But none of the nasal swabs allowed influenza virus isolation.

Fourteen among the seventy-seven samples tested against <u>Dermatophilus</u> <u>congolensis</u> gave a positive reaction. Nevertheless no gross cutaneous lesion was observed during this period and up to now. So the contact between Red deer and pathogen was that considered as certain but either the animal resisted, either the lesions healed spontaneously.

A similar explanation is possible concerning the positive reaction against <u>Foot</u> <u>and</u> <u>mouth</u> <u>disease,</u> C type which was found during the first captures. Afterwards 77 probang tests were performed, but isolation of FMD virus was unsuccessful.

No clinical signs related to <u>paratuberculosis,</u> <u>Q</u> <u>fever</u> and <u>chlamydiosis</u> were observed.

<u>Brucellosis</u> serology was performed in two steps : first a screening with "rose Bengale test" then the positives selected by this test were controlled by a complement fixation reaction. Two among fifty-four animals were positive to both reactions. Serological analysis being unable to demonstrate a difference between the different species of <u>Brucella,</u> it is strongly suggested that these two animals were contaminated indirectly from domestic animals (cattle unnoticed brucellic abortion around).

Some reactions were positive against three <u>arboviruses</u> (one phlebovirus : Sandfly Fever Sicile type, one orbivirus : Eyach and a new virus isolated in 1982, the "Brest/AN 212 virus") (Le Lay-Rogues and coll., 1987). The first of these three viruses had never been isolated in continental France so far. It is nevertheless present in Corsica where its vector is known to be <u>Phlebotomus</u> <u>papatasi</u> : this insect may also be found in the Mediterranean region of France only. So one can suggest that other <u>Phlebotomus</u> sp. may be vectors for this virus, in other parts of France.

The results concerning Herpes virus infection will be detailed by Dr. Thiry during this seminar.

DISCUSSION-CONCLUSION

Serological survey of wildlife is of uppermost interest in comparative epidemiology of different pathogens, but the accuracy of the results is often more questionable than in domestic animals. One of the reasons is that controlling the kinetics of antibody response is impossible. Either the animals are captured and then released and never trapped again (it happened only twice, among near 1000 wild ungulates) or they are killed by hunters.

The second difference with domestic animals is that positive thresholds are not really known and are merely extrapolated from which is known in domestic animals.

Actually despite these two restrictions, the positive reactions mean that at least one contact have occurred between the antigen and the animal. And results obtained during our study are closely comparable to those obtained with other animal species in France, and abroad, where severe disease incidence were never reported.

If these results show that the Red deer is not currently affected by diseases commonly recognised in domestic ruminants in France, they show that this species, is, nevertheless, susceptible to contamination by various pathogens. In other conditions (e.g. farming or breeding of the species in closed areas) this equilibrium between the pathogens and their hosts could be destroyed, and severe diseases emerge in these populations. Natural selection observed in free-living animals will in this case, play no role in elimination of susceptible animals.

ACKNOWLEDGMENTS
    We wish to thank here the laboratories that collaborated to this study :
- Faculté de Médecine de Brest, B.P. 815, 29700 Brest Cédex, France (Cl. Chastel)
- Ministère de l'Agriculture, Direction Générale de l'Alimentation, Services Vétérinaires, Laboratoire National de Pathologie Bovine, B.P. 7033, 69342 Lyon Cédex, France (G. Dannacher).
- Ministère de l'Agriculture, Direction Générale de l'Alimentation, Services Vétérinaires, Laboratoire Central de Recherches Vétérinaires, B.P. 67, 94703 Maisons-Alfort Cédex, France (J.M. Gourreau, B. Larenaudie, D. Trap)
- Institut d'Elevage et de Médecine Vétérinaire des Pays Tropicaux, 10 rue Pierre Curie, 94704 Maisons-Alfort Cédex, France (C. Le Goff, P. Perreau).
- Laboratoire Vétérinaire Départemental de Meurthe et Moselle, B.P. 39 , 54220 Malzéville, France (J.M. Baradel)
- Office National de la Chasse, 109 rue St Georges, 54000 Nancy, France (F. Klein, B. Boisaubert, J. Vincent).
- Office National des Forêts, Avenue Ashton Underline, 52000 Chaumont, France (Y. Leprince, J. Charbonnel)

REFERENCES

Blancou J., 1983. Serologic testing of wild Roe deer (Capreolus capreolus)
Blancou, J. 1984. Pathologie des ongulés sauvages de France. Bilan des recherches récentes. Gibier Faune Sauvage, 87-95.
Le Lay-Rogues, G., Barrat, J., Hardy Le Maux, E., Aubert, M., et Chastel, C. 1987. Infections à arbovirus en France : enquête sérologique chez les grands mammifères sauvages. Méd. Mal. Infect., 370-376.

ETHOLOGICAL ASPECTS OF DEER FARMING

H. Hemmer

Institute of Zoology, Johannes Gutenberg University,
Saarstrasse 21, D-6500 Mainz, F. R. Germany

ABSTRACT

The species-specific behaviour regulation of environmental relations governs the qualification of European deer species to be used in economically successful game farming programs. New results on this topic are reviewed. Qualitative as well as quantitative aspects of social behaviour point to a first class qualification of the fallow deer (Dama dama), a moderate qualification of the sika deer (Cervus nippon) and the red deer (Cervus elaphus) and a lack of qualification in the elk (Alces alces) and the roe deer (Capreolus capreolus). The emotional sensitivity making roe deer farming nearly impossible is causing much trouble in fallow deer farming, sika deer follows, red deer and especially elk are easier to manipulate. Aggression in hand-raised males is highest in roe deer, but fallow deer obviously just follows. Red deer causes less problems. On the contrary, male sika deer and red deer may be extremely aggressive to strangers. These features govern constructive needs in enclosure size, shape and structure and in fencing. Altogether fallow deer can be reshaped to the incontestable top position for deer farming purposes if its special panic problems are damped by the genetically selective process of domestication. The great behavioural advantage of domestic fallow deer in the whole management and handling context is reviewed by introducing the first true domestic fallow deer founder group.

INTRODUCTION

Behaviour is the way of animals to regulate their environmental relations. Like all organismic regulation mechanisms this is an energy costing procedure. Evolution specifically stabilizes the adequate behavioural energy budget to regulate daily life comfortably in the specific environment at a level that usually will not endanger the total energy balance. Artificial change of this natural environment by farming an animal within an enclosure should have the greater impact on this behaviour energy budget the more the animal is pressed under unfavourable conditions where it is forced by its specific central nervous system innate as well as experienced programs to a continual counterregulation that nevertheless finally never can be successful to terminate the lasting trouble. Chronically enhanced stress levels must be the result if this balance is continuously overridden by an unfit design of the captive habitat. This opens into much reduced individual general fitness with enhanced susceptibility to parasitism and infection, causing artificially created health problems in farmed game animals that could be greatly reduced from the first if ethological aspects would be better noticed in the choice of the species to be farmed and of the enclosure type to be used.

A shortage of food resources surely will not be a common problem in deer farming that drives the behavioural energy balance out of regularity. Also predator pressure will not be such a problem in farms. Really large-sized enclosures with an animal density not exceeding the mean natural density should not result in any enhanced stress level, if there is no continuous irritation by the farmer. This type of enclosure allows all species to be farmed independently of their specific sensibility to environmental influences. But such a farming system seldom will have an economic value. The more energetic trouble will be caused at the other hand if the basic safety requirements are continuously stressed by too small an area or by inevitable continual anthropogenic irritations, or if the regulative mechanisms to maintain social stability are continuously overstressed by too high a population density or by frequent forced changes in group composition.

The process of true domestication has been the measure of man through the millenia to adapt and shape certain preadapted animal species genetically by behavioural selection into such organisms that react much less sensitive to all pertinent problems in an artificial farming environment, being buffered by their typically impoverished perceived world (Hemmer, 1983) when being compared with their original wild relatives. Therefore it should be wise to look for specific behavioural features of true domestic animals as an outside comparative unit better to evaluate the specific qualification of different wild game species taken into consideration for farming purposes.

SOCIAL BEHAVIOUR AND SPECIES' QUALIFICATION
The work of Rammelsberg (1986 c) that surely will stand as a basic source book for a long time did this type of compa-

rative evaluation for the social behaviour of European deer
species. A summary of her results in view of specific qualifi-
cations for deer farming (Rammelsberg 1986 a) should shortly
be reviewed here. Highly elaborate differentiation of patterns
of social behaviour, high behavioural intensities, high emotional participation in social interactions, the maintainance
of highly complex social systems or the need always to re-es-
tablish unstable systems will cost high amounts of behavioural
energy in deer farms, if the farmed animal groups cannot stay
regulated on a stabilized state and with a clearly adequate
number of specimens. Just this looks to be impossible in small
or medium-sized deer farms (types 2 and 3 farms as discussed
by Hemmer, 1984) continuously operated in an economic view-
point, changing the group composition e.g. by seasonally different management measures, by breeding selection, by slaughter.
Looking on such central qualitative criteria of social behaviour patterns presents fallow deer (Dama dama) as the most
qualified species, sika deer (Cervus nippon) and red deer (Cervus elaphus) following as moderately suited ones, elk (Alces alces) looks to be the most unsuited species. Roe deer (Capreolus capreolus) is ranking relatively high in the order, if
simple rank numbers are counted, but its really very high behavioural intensity must seriously put this place in question
(table 1).

The use of purely quantitative criteria as being of special interest for herd management, i.e. the influence of season and the influence of changed captive habitat, the level of
socially relevant behavioural intensities, the general social
activity and especially the basic social tolerance, allows a
much similar species arrangement (Rammelsberg 1986 a / table 2)
where fallow deer takes the top position, followed by sika
deer and red deer, then the elk and finally roe deer.

These results prove altogether that social behaviour parameters allow fallow deer a balanced life under continuously
changed and mostly crowded conditions in deer farming much
better than any other European deer species. Sika deer and red
deer show only moderate qulifications compared to fallow deer,
elk and roe deer are unsuited.

TABLE 1   Specific qualification of European deer species to become domesticated: qualitative criteria of social behaviour. Basic classification as done by Rammelsberg (1986 a), rank numbers newly added.

	Roe Deer	Fallow Deer	Sika Deer	Red Deer	Elk
Social behaviour pattern differentiat.	low 4	moderate 3	strong 2	strong 2	very strong 1
Behavioural intensity	very high 1	low 4	moderate 3	high 2	high 2
Emotional participation in soc.interact.	moderate 3	low 4	high 2	high 2	very high 1
Complexity of group structure	no stable rank order 4	simple rank order 3	simple rank order 3	complex rank order 2	complex relationships 1
Qualification to become domesticated	(moderate) 12	high 14	moderate 10	moderate 8	low 5

TABLE 2   Specific qualification of European deer species to become domesticated: rank order by quantitative criteria. Basic classification as done by Rammelsberg (1986 a), rank numbers newly added.

	Roe Deer	Fallow Deer	Sika Deer	Red Deer	Elk
Low level of soc.activity	+	+++++	+++	++++	++
Low frequency of aggression	+	+++++	++++	+++	++
Low social intolerance	+	+++++	++++	++	+++
Low influence of capt.habit.	+	+++	++	+	+
Low influence of season/♀♀	++++	+++++	+++	++	+
Low behavioural intensity	+	+++	++	++	++
Qualification to become domesticated	low 9	high 26	moderate 18	moderate 14	low 11

EMOTIONALITY, AGGRESSION AND SPECIES' QUALIFICATION

Safety requirements of the different species may be estimated by measuring escape distances, reactivity to external
stimulation or by quantification of postural signs of excitement. Comparative calculations of this type done with the same
methods are still lacking. But there are general comparative
impressions of the emotional sensitivity of the European deer
species by experienced deer farming or zoo people. The roe
deer looks as the most problematical one. Not only its captive
breeding but even its long time keeping often proves to be extremely difficult. The fallow deer follows in its notorious
readiness to take flight and to panic. Sika deer may be arranged somewhere between fallow deer and red deer. Red deer are
said to be much easier to manipulate in captivity, even being
more distrustful towards man at first (Schulz 1986 a). Elk is
still less ready to take flight than red deer (Schulz in Jung
1986 a). This species arrangement follows the line of progressive cephalization in the compared species. Relative brain size measurements indeed set the roe deer at the lower end of
the scale, followed after some gap by fallow deer and sika
deer, then red deer and finally elk on the top. This suggests
the interpretation that managing the species in game farms in
view of their emotional reactivity should be the easier the
higher is the cephalization level, i.e. the better is the capability of the species in question to acquire elaborate learned behavioural programs related not only to the complexity of
social systems but also to the attitude towards continuous anthropogenic stimulations and irritations.

Aggressive interactions of farmed deer, especially males
during rut, with man also may correlate to some extent with
cephalization level, but also with body size. Hand-reared male
roe deer is widely known for its highly aggressive nature.
Hand-reared male fallow deer obviously showns similar aggressiveness in some individuals as the author experienced in the
Riswick program (Germany). Male red deer seem to be more reli-
able for the attending and intimate persons under the same conditions. On the contrary, red deer, even females, must be
classified as highly dangerous towards strangers, and sika

deer males are also said to be absolutely aggressive. Not hand-
reared male fallow deer shows very low aggression when being in
contact with man (e.g. Blaxter et al, 1974, Bogner in Jung,1986
a, Hemmer,1986 a, Pietrowski, 1984, Rammelsberg, 1986 a,Schulz,
1986 a). Therefore hand-rearing may be a helpful technique for
easy handling red deer and elk by special people, but as this
cannot be the rule in commercially operated deer farms, red
deer has a practically highly important disadvantage for all
handling procedures, especially during rut. Sika deer looks not
to be much better. Fallow deer allows easy handling in view of
low aggressiveness, but hand-rearing should be strictly avoided in this species other than for scientific purposes, not to
produce artificially aggression problems.

The safety-proved stability of fences, but also their
height, must correspond to this specific attitude of different
deer species towards man under the regimen of captivity. Red
deer and sika deer force much greater expenses in this point
than fallow deer can do. The immense specific farming problems
with fallow deer on the contrary result by their inclination
to panic. Therefore this species should be provided in view of
economic success as well as of animal welfare with especially
great depths of the pasture, allowing at least a retreat to a
flight distance march of at least 50 m (e.g. Pietrowski, 1984)
from public ways in all such farms that do not have daily intense human contacts as being the case in zoo enclosures where
accustomation to continuous irritations is possible. Highly
structured enclosures also help to compensate for this special
specific behavioural disadvantage in fallow deer (e.g. Bamberg,
1985, 1986, Hemmer in Klein 1986, Schulz 1986 b).

BEHAVIOURAL CHANGES BY DOMESTICATION

As the previous chapters did show, fallow deer is clearly
better qualified for deer farming (as stated especially by
Reinken since a decade: 1977, 1980 etc.) than all other European species in view of its social behaviour and its low
aggressiveness. The only but really important disadvantage of
this species for captive management is the characteristic panic problem. It must be reduced to reshape fallow deer for the

incontestable top position in deer farming. This can be reached
by the domestication process. It must be stressed here that the
term domestication is very often misused in the sense of mere
taming an animal (e.g. Blaxter et al., 1974). There is also sometimes made the sophistic difference in the English language
between domestic and domesticated animals, both qualities resulting from domestication what so ever this then may be. To get
a clear and sound basis fur further discussion in this field,
we should be aware that the difference between domestication
and taming is just nothing else than the really basic difference between phylogeny and ontogeny in animal development. True
domestication means getting a new behavioural design by changing the genetic basis of information acquisition and information processing (Hemmer, 1983). Taming on the other hand means
nothing else than adapting an animal by its individual experience to the regimen of humans, to tolerate close human presence. Therefore taming may be the easier the higher cephalization level is found in a species in question.

   There may be distinguished between three basic types of
tameness, i.e.:

   (1) tameness by primary acquisition of confidence as done
by hand-rearing,

   (2) tameness by reduce of distrust, as done in taming
adult animals,

   (3) tameness by naivety as the only type of tameness on
a genetic basis. Only this last one is the basic type of tameness as typical for real domestic animals.

   The difference may be made plain in looking on the reaction of hand-reared (Riswick program) and domestic but not especially tamed (Neumühle program, transferred to Riswick)
young male fallow deer acting side by side. Attracting them
to the fence by a person not being familiar with these individuals resulted (reproduced for several times) in mere neutral
action of the one year old domestic male but in intense con-
tacting by the hand-reared two years old males. Then suddenly
being frightened by an unexpectable action of the observer,
the hand-reared animals stay nearly, but show all postural
signs of being alarmed, whilst the domestic animal still re-

mains in a much more neutral position. The last step shows one
of the hand-reared males attacking, the domestic one still
neutral. Just this is the difference between a fully tame, but
emotionally participating, genetically unchanged wild deer and
a somewhat less tame but genetically reshaped true domestic
animal.

Historically tha domestication of large mammalian species
as the classic domestic ones like cattle, sheep, goat etc. surely took long time, may be more than a millenium, probably
not being processed by known strategies but by chance development on the basis of game farming enterprises lasting for centuries. Having the know-how for projected domestication (Hemmer, 1983), this process now may be reached in a few animal generations only, beginning with the strict selection of preadapted individuals as founder stock and then working with methods of outsider selection providing greatest heritability
and of basic feature combination. Just such a way has been gone
in the fallow deer domestication program runned at the Neumühle institute (Germany) since 1979 and resulting within less
than a decade in obviously the first true domestic fallow deer
(Gaede, 1986, Hemmer, 1986 b, c, Klein, 1986 b, Rammelsberg
1986 b).

There was never hand-rearing or even any continuous con-
scious attempt of taming at all in this program. Nevertheless
the behavioural results of this genetical redesigning fallow
deer are obvious. The primary main point is the reduction of
shyness and frightfulness to an unproblematical measure that
facilitates handling in such a way as to be awaited for unimproved breeds of classic domestic species. The Neumühle domestic fallow deer stock retains the behavioural advantages of
wild fallow deer compared to the other European species, even
enhancing the basic positive traits, and combines them with
good tolerance of being handled without the need of any intense taming efforts, and also with the qualification to be farmed in much less well structured enclosures as necessary for
genetically unchanged deer.

The desired animal that combines all economic advantages
as farmed deer may provide with the basic ethological charac-

teristics of unimproved breeds of classic domestic herbivore
species - here it is for further propagation and use in the
European Community and elsewhere.

REFERENCES

Bamberg, F. 1985. Untersuchungen an gefangenschaftsbedingten
    Verhaltensänderungen beim Damwild (Cervus dama Linne,
    1758). Beitr. z. Wildbiol. 5. (Hartmann, Kiel).
Bamberg, F. 1986. 2.2.1 in "Nutztier Damhirsch" (Ed. H. Hem-
    mer). (Rhein.Landwirtsch.verlag, Bonn). pp.42-47.
Blaxter, K.L., Kay, R.N.B., Sharman, G.A.M., Cunningham, J.M.M.
    and Hamilton, W.J. 1974. Farming the Red Deer. (Depart-
    ment of Agriculture and Fisheries for Sootland, Edinburgh)
Gaede, E.A. 1986. 4.2.1 in "Nutztier Damhirsch" (Ed. H. Hemmer)
    (Rhein.Landwirtsch.verlag, Bonn).pp. 95-112.
Hemmer, H. 1983. Domestikation - Verarmung der Merkwelt. (Vie-
    weg, Braunschweig/Wiesbaden).
Hemmer, H. 1984. The aptitude and selection of large mammals to
    game farming and domestication. Acta zool.Fennica, 172,
    233-236.
Hemmer, H. 1986 a. 1.2.1 in "Nutztier Damhirsch" (Ed. H. Hem-
    mer). (Rhein.Landwirtsch.verlag, Bonn). pp. 8-11.
Hemmer, H. 1986 b. 3.2.3 in "Nutztier Damhirsch" (Ed. H. Hem-
    mer). (Rhein.Landwirtsch.verlag, Bonn). pp. 82-91.
Hemmer, H. 1986 c. 4.2.2 in "Nutztier Damhirsch" (Ed. H. Hem-
    mer). (Rhein.Landwirtsch.verlag, Bonn). pp.112-117.
Jung, P. 1986 a. 1.3 in "Nutztier Damhirsch" (Ed. H. Hemmer).
    (Rhein.Landwirtsch.verlag, Bonn). pp. 38-40.
Jung, P. 1986 b. 3.2.1 in "Nutztier Damhirsch" (Ed. H. Hemmer).
    (Rhein.Landwirtsch.verlag, Bonn). pp. 71-77.
Klein, H. 1986 a. 2.3 in "Nutztier Damhirsch" (Ed. H. Hemmer).
    (Rhein.Landwirtsch.verlag, Bonn). pp. 64-69.
Klein, H. 1986 b. 3.3 in "Nutztier Damhirsch" (Ed. H. Hemmer).
    (Rhein.Landwirtsch.verlag, Bonn). pp. 92-93.
Pietrowski,K. 1984. Untersuchungen zum Verhalten von Damwild
    bei nutztierartiger Haltung. Ph.D.Thesis (Bonn).
Rammelsberg, C. 1986 a. 1.2.2 in "Nutztier Damhirsch". (Ed. H.
    Hemmer). (Rhein.Landwirtsch.verlag, Bonn). pp. 11-32.
Rammelsberg, C. 1986 b. 3.2.2 in "Nutztier Damhirsch" (Ed. H.
    Hemmer). (Rhein.Landwirtsch.verlag, Bonn). pp. 78-82.
Rammelsberg, C. 1986 c. Vergleichende Studien zum Sozialverhal-
    ten europäischer Cerviden. Ph.D.Thesis (Mainz).
Reinken, G. 1977. Grün- und Brachlandnutzung durch Damtiere.
    (Landwirtschaftskammer, Bonn).
Reinken, G. 1980. Damtierhaltung auf Grün- und Brachland. (Ul-
    mer, Stuttgart).
Reinken, G. 1986 a. 1.2.4 in "Nutztier Damhirsch" (Ed. H. Hem-
    mer). (Rhein.Landwirtsch.verlag, Bonn). pp. 37-38.
Reinken, G. 1986 b. 2.2.2 in "Nutztier Damhirsch". (Ed. H. Hem-
    mer). (Rhein.Landwirtsch.verlag, Bonn). pp. 48-49.
Schulz, G. 1986 a. 1.2.3 in "Nutztier Damhirsch". (Ed. H. Hem-
    mer). (Rhein.Landwirtsch.verlag, Bonn). pp. 32-36.
Schulz, G. 1986 b. 2.2.3 in "Nutztier Damhirsch". (Ed. H. Hem-
    mer). (Rhein.Landwirtsch.verlag, Bonn). pp. 49-53.

Conclusions from Session III

1. Health aspects of deer farming

Preventive medicine was important on economic and ethical grounds. One condition was copper poisoning in fallow deer which appeared to have a susceptibility similar to sheep and therefore fallow deer should be fed copper at a similar rate to sheep. However in discussion it was pointed out that red deer fed 50 ppm of copper appeared to demonstrate a better growth rate. The speaker then asserted that necrobacillosis due to trauma of the buccal mucosa could be caused by feeding coarse hay.

2. Tuberculosis in farmed deer

Discussion centred around the problem of diagnosing TB in deer as both the skin test and serological methods of diagnosis were inadequate. In view of this the speaker agreed with one questioner that testing for in-vitro lymphocyte activation would be well worth investigating as a means of diagnosis even though it had proved problematical in other disease systems.

The urgency for improved diagnostic methods for the mycobacteria was emphasised by the speaker when she pointed out that TB in deer can spread quickly and that at post mortem examination it does not resemble TB of cattle. TB in deer can produce large abscesses filled with pus. All such lesions must be viewed as potentially due to TB.

3. Paratuberculosis (Johne's disease)

During the paper the speaker indicated that, unlike paratuberculosis in cattle, in deer the intestines and mesenteric lymph nodes may appear relatively normal macroscopically, lesions only being visible histologically. It was also stated that deer do not respond to mycobacteria in the same way that cattle and sheep do. In reinforcing the previous speaker's comments on the diagnosis of TB it was pointed out that no practical diagnostic test for paratuberculosis of deer existed other than the slow and difficult method of faecal culture.

During the discussion it was concluded that M. avium and M. paratuberculosis could present very similar pathology in deer and that current methods of distinguishing between the two were inadequate. It

139

was concluded that paratuberculosis of deer was potentially very serious for the industry. The speaker agreed that without more research the outlook was bleak.

4. Serological survey of free-living red deer (_Cervus elaphus_) in France

During the presentation the speaker gave serological evidence of several infections being present in the deer population sampled. One of these was paratuberculosis and the discussion commenced with this condition. The speaker said that no clinical disease had been found but as the CFT was not sufficiently specific other mycobacterial infections may be present.

5. Ethological aspects with regard to deer farming

During the discussion the speaker confirmed that he had domesticated, rather than tamed, fallow deer after selective breeding for two generations. He stated that the domesticated animals were heavier than wild fallow deer and had altered coat colours. It was concluded that red and sika deer could also be domesticated by selective breeding.

<u>SESSION IV</u>

Chairman: Professor P-P. Pastoret

Co-Chairman: Dr. H.W. Reid

STUDIES ON THE EPIDEMIOLOGY AND PATHOGENESIS OF ALPHAHERPESVIRUSES FROM RED DEER (<u>CERVUS ELAPHUS</u>) AND REINDEER (<u>RANGIFER TARANDUS</u>)

P.F. Nettleton[1], C. Ek-Kommonen[2], R. Tanskanen[3], H.W. Reid[1], J.A. Sinclair[1], J.A. Herring[1]

[1]Moredun Research Institute, 408 Gilmerton Road, Edinburgh, EH17 7JH Scotland
[2]National Veterinary Institute, Hameentie 57, 00550 Helsinki 55, Finland
[3]Department of Microbiology and Epizootology, Veterinary College of Helsinki, Finland

ABSTRACT

Alphaherpesviruses, serologically related to bovine herpes virus 1 (BHV-1) have been isolated from red deer (<u>Cervus</u> <u>elaphus</u>) and reindeer (<u>Rangifer</u> <u>tarandus</u>). Biochemical and serological studies have demonstrated that these two viruses are distinct from each other and from BHV-1.

The red deer herpesvirus, tentatively designated herpesvirus of Cervidae type 1 (HVC-1) was recovered in 1982 from red deer calves with ocular disease in the North of Scotland and a further seven outbreaks of disease occurred in 1983. A serological survey has shown that the virus is widespread in free-living and farmed red deer. Cattle experimentally infected with HVC-1 showed no evidence of disease.

In Finland, surveys have shown that cattle do not have serological evidence of infection with BHV-1. On the contrary, approximately 20% of reindeer in Finnish Lapland have antibody to BHV-1. The reindeer herpesvirus was isolated from the vagina of one seropositive hind which had been treated with dexamethasone. The experimental infection of 4 cattle with the virus showed that it was mildly pathogenic in this species.

INTRODUCTION

Antibody to the cattle alphaherpesvirus bovine herpesvirus type-1 (BHV-1), the causative agent of infectious bovine rhinotracheitis and infectious pustular vulvovaginitis, has been detected in 5 serological surveys of deer. Mule deer (<u>Odocoileus</u> <u>hemionus</u> <u>hemionus</u>), white-tailed deer (<u>O.</u> <u>virginianus</u>) and caribou (<u>Rangifer</u> <u>tarandus</u> <u>arcticus</u>) were surveyed in North America (Chow and Davies, 1964; Friend and Halterman, 1967; El Azhary, 1979) while red deer (<u>Cervus</u> <u>elaphus</u>) and reindeer (<u>Rangifer</u> <u>tarandus</u>) were tested in Scotland and Finland respectively (Lawman et al., 1978; Ek-Kommonen, et al., 1982).

143

In 1982, an alphaherpesvirus serologically related to BHV-1 was isolated from pooled ocular/nasal swabs from red deer suffering from eye disease (Inglis et al, 1983). The same virus was recovered from further outbreaks of eye disease in 1983 (Nettleton et al, 1986) but no disease due to the virus has been recorded in Scotland since then. An alphaherpesvirus was also recently isolated from the vagina of a reindeeer treated experimentally with dexamethasone (Ek-Kommonen et al., 1986). However no disease attributable to alphaherpesvirus infection has been observed in reindeer. Restriction endonuclease analysis of DNA from these two viruses has shown that they are distinct from each other and from BHV-1 (A.J. Herring, unpublished; Ronsholt et al., 1987). The red deer herpesvirus has been tentatively designated herpesvirus of cervidae type 1 (HVC-1) (Reid, et al., 1986).

The purpose of this paper is to summarise and compare the results of experimental infection of cattle with each of the two viruses and their prevalence of infection in the two species of deer in natural populations.

MATERIALS AND METHODS

Viruses

Low passage levels of both viruses grown in embryonic bovine trachea or turbinate cells were used to infect the experimental cattle. The titre of the HVC-1 virus was 5×10^5 tissue culture infective doses 50 ($TC1D_{50}$) per ml and that of the reindeer virus, which will be referred to as herpesvirus of cervidae type 2 (HVC-2), was 6×10^6 $TCID_{50}$ per ml.

Experimental animals

Cattle receiving HVC-1. As reported in detail by Reid et al., (1986) two jersey calves negative for antibody to BHV-1 and HVC-1 were both inoculated with 1 ml of virus in each nostril. Eighty-four days later the calves were challenged intranasally with 5×10^7 $TCID_{50}$ of a virulent field isolate of BHV-1.

The cattle were examined clinically and their rectal temperatures recorded daily for 10 days after each inoculation. Nasal and ocular swabs were collected for virus isolation during these periods. Sera were collected before inoculation and on days 7 and 14 after exposure for measurement of antibody to HVC-1 and BHV-1.

Cattle receiving HVC-2. This experiment will be reported in full elsewhere (Ek-Kommonen et al, in preparation). Briefly, 4 cattle, which were free of antibody to BHV-1 and HVC-2, were infected by different routes. Every day, for two weeks after infection, animals were examined clinically and nasal and vaginal swabs were collected for virus isolation. Sera were collected twice a week for 10 weeks for antibody measurement.

Serology

Microneutralisation tests using approximately 40-50 $TCID_{50}$ of virus/well were employed to measure serum antibody to the two deer viruses and to BHV-1. Before the addition of cells, mixtures of virus and serum from cattle infected with HVC-1 or HVC-2 were held at $37^{o}C$ for 1 hour and 24 hours respectively.

Serological surveys

Sera to be tested for antibody to HVC-1 were collected from 520 red deer from three different sources. 145 culled Scottish hill deer, 203 Scottish farmed deer and 172 English farmed or park deer. The deer could be classified as adults (> 1 year old) or calves (< 1 year old) (Nettleton et al, 1986).

Sera tested against the reindeer herpesvirus were collected from 193 adult reindeer older than 18 months and 274 five to nine month old reindeer calves in North Finland (Ek-Kommonen et al, in preparation).

All sera were tested against their respective viruses in microneutralisation tests as described above. A reciprocal serum dilution of 2 or more was taken as positive evidence of infection.

RESULTS
Experimental infection of cattle

Cattle receiving HVC-1. No pyrexia or clinical disease occurred in the two calves after infection with HVC-1. No virus was ever recovered from one of the calves while virus was recovered from the nasal swabs from one calf collected on days one and two after infection. Both calves developed low antibody titres to HVC-1 and to BHV-1 virus detectable 7 days after infection. When challenged with BHV-1 84 days after the infection with HVC-1, both calves proved fully susceptible to the challenge: both developed pyrexia and nasal discharges and virus was recovered in nasal and ocular swabs for 7 days after infection.

Cattle receiving HVC-2. Although mild symptoms of rhinitis was observed none of the animals showed any signs of serious disease and there was no loss of appetite. Virus was isolated for 6 to 9 days from nasal and/or vaginal swabs of all the cattle, and all animals developed antibody to the virus within 8 to 12 days of infection. When the sera were tested against BHV-1 the reciprocal titres were 2 fold lower than against the homologous virus.

Serological surveys

Of the 520 red deer sera tested, 149 (29%) contained antibody to HVC-1. The highest prevalence (40%) was in Scottish hill deer followed by Scottish farmed deer (33%) and English deer (14%). One hundred and forty one of 408 (35%) sera from adults were positive compared to 8 of 112 (7%) sera from calves. Among adult Scottish hill deer 57 of 139 (41%) animals were positive.

Of the 467 reindeer sera tested 146 (31%) contained antibody to HVC-2. The prevalence of antibody was very high among the adult animals with 124 (64%) being positive, while only 18 (1%) of the sera from the calves contained antibody.

DISCUSSION

The serological surveys have shown that both viruses are well established in their own host species. The viruses have clearly evolved in natural populations of their host and, given the prevalence of antibody among free-living adults as 64% for reindeer and 41% for red deer, they are spread efficiently without ever causing serious disease. No natural disease has been observed in reindeer while disease in red deer has only been observed in calves following the stress of weaning, housing and transport.

The mechanisms of viral spread in natural populations of the two species are not known, but analogy with the closely related alphaherpesvirus of cattle, BHV-1, make it likely that both the respiratory and genital routes are important (Gibbs & Rweyemamu, 1977). The evidence available so far is that the genital route is the most important means of spread among reindeer, while the ocular disease seen in red deer suggests the respiratory route is more important in this species. The most likely explanation however, is that either route of infection may occur in either species.

Experimental infections of cattle with HVC-1 showed that no disease and only minimal viral replication occurred. A similar result has recently been reported by Ronsholt et al, (1987) who also demonstrated that cattle kept in close contact with deer excreting HVC-1 did not become infected by the virus. Natural infection of cattle with HVC-1 from red deer must therefore be considered almost impossible.

Under experimental conditions, cattle infected with HVC-2 did not show any signs of serious disease, but virus was readily isolated from them and they seroconverted suggesting that this agent can replicate in cattle. There is however, no evidence of natural transmission occurring. Cattle in Finland are screened continuously for the presence of antibody to BHV-1 and no seropositive animals have yet been detected even in areas where both cattle and reindeer are present (Ek-Kommonen et al., 1982). We would agree with the conclusion of Brake and Studdert (1985) that each of the ruminant alphaherpesviruses has a restricted epidemiology and that under natural conditions the viruses do not cross species barriers.

Nomenclature of these viruses remains a problem. Since the red deer virus was the first alphaherpesvirus to be identified as a unique cervid virus the designation cervid herpesvirus type 1 (CHV-1) has been proposed (Ronsholt et al., 1987). Anticipating that this acronym would further confuse that already used for canine herpesvirus type 1 and caprine herpesvirus type 1 Reid et al. (1986) proposed herpesvirus of cervidae type 1 (HVC-1). The acronym CerHV-1 is a third possibility for consideration. Whatever the final decision it is clear that distinctions between the viruses will need to be recognised and they could logically be termed HVC-1 or CerHV-1 for the red deer virus and HVC-2 or CerHV-2 for the reindeer virus. Until other herpesviruses are isolated from these species, however, the current isolates will be readily identifiable as red deer herpesvirus and reindeer herpesvirus, respectively.

Addendum

Since writing this paper it has been drawn to our attention that clinical ocular disease similar to that previously reported has been observed in 1987 in market purchased red deer calves (J. Fletcher, personal communication).

REFERENCES

Brake, F. and Studdert, M.J. 1985. Molecular epidemiology and
     pathogenesis of ruminant herpesviruses including bovine,
     buffalo and caprine herpesviruses 1 and bovine encephalitis
     herpesvirus. Austr. Vet. J. 62, 331-334.
Chow, T.L. and Davies, R.W. 1964. The susceptibility of mule deer
     to infectious bovine rhinotracheitis. Amer. J. Vet. Res. 25,
     518-519.
Ek-Kommonen, C., Pelkonen, S. and Nettleton, P.F. 1986. Isolation of a
     herpesvirus serologically related to bovine herpesvirus 1 from a
     reindeer (Rangifer tarandus). Acta. Vet. Scand. 27, 299-301.
Ek-Kommonen, C., Veijalainen, P., Rantala, M. and Neuvonen, E. 1982.
     Neutralising antibodies to bovine herpesvirus 1 in reindeer.
     Acta Vet. Scand. 23, 565-569
El-Azhary, S. 1979. Serological evidence of IBR and BVD infection in
     caribou (Rangifer tarandus). Vet. Rec. 105, 336.
Friend, M. and Halterman, L.G. 1967. Serological survey of two deer
     herds in New York State. Bull. Wildl. Dis. Ass. 3, 32-34.
Gibbs, E.P.J. and Rweyemamu, M. 1977. Bovine herpesviruses. Part 1.
     Bovine herpesvirus 1. Vet. Bull. 47, 317-343.
Inglis, D.M., Bowie, J.M., Allan, M.J. and Nettleton, P.F. 1983.
     Ocular disease in red deer calves associated with a herpesvirus
     infection. Vet. Rec. 113, 182-183.
Lawman, M.J.P., Evans, D., Gibbs, E.P.J., McDiarmid, A. and Rowe, L.
     1978. A preliminary survey of British deer for antibody to some
     virus diseases of farm animals. Br. Vet. J. 134, 85-91.
Nettleton, P.F., Sinclair, J.A., Herring, J.A., Inglis, D.M., Fletcher,
     T.J., Ross, H.M. and Bonniwell, M.A. 1986. Prevalence of
     herpesvirus infection in British red deer and investigation of
     further disease outbreaks. Vet. Rec. 118, 267-270.
Reid, H.W., Nettleton, P.F., Pow, I. and Sinclair, J. 1986. Experimental
     infection of red deer and cattle with a herpesvirus isolated from
     red deer. Vet. Rec. 118, 156-158.
Ronsholt, L., Siig Christensen, L. and Bitsch, V. 1987. Latent
     herpesvirus infection in red deer: characterisation of a specific
     deer herpesvirus including comparison of genomic restriction
     fragment patterns. Acta. vet. Scand. 28, 23-31.

HERPESVIRUS INFECTIONS OF RED DEER AND OTHER WILD RUMINANTS IN FRANCE AND BELGIUM

E. Thiry*, P.-P. Pastoret*, J. Barrat***, J. Dubuisson*,
J. Blancou***, B. Collin**.

* Department of Virology and ** Centre de Médecine du Gibier
Faculty of Veterinary Medicine, University of Liège,
B-1070 Brussels, Belgium
*** Centre National d'Etudes sur la Rage et la
Pathologie des Animaux Sauvages,
F-54220 Malzéville, France.

ABSTRACT

The epizootiology of herpesvirus infections in wild ruminants of France and Belgium is reviewed, with particular reference to red deer (Cervus elaphus). Infections by the following viruses are presented : bovine herpesvirus 1 (BHV-1), bovine herpesvirus 2 (BHV-2), bovid herpesvirus 4 (BHV-4), caprine herpesvirus 2 (CHV-2), reindeer (Rangifer tarandus) herpesvirus, herpesvirus of Cervidae type 1 (HVC-1). A serological survey was undertaken between 1981 and 1986. No animals were seropositive against BHV-4. The prevalence of antibodies against BHV-2 was very low : <1% in roe deer (Capreolus capreolus), 1% in chamois (Rupicapra rupicapra). No antibodies against BHV-2 were demonstrated in red deer. Four herpesviruses are serologically related: BHV-1, CHV-2, reindeer herpesvirus and HVC-1. Antibodies against one or several of these viruses were detected in the following species: chamois (4%), ibex (Capra ibex) (4%), roe deer (<1%); the prevalence of red deer seropositive against one of the 4 related viruses was different in France (1%) and in Belgium (11%). The geographical distribution of seropositive animals is discussed as well as the probable presence of HVC-1 in red deer in Belgium. The results obtained between 1981 and 1986 are compared to those obtained in Great Britain and in other European countries.

INTRODUCTION

Herpesvirus infections of wild ruminants have not yet been associated with significant losses in European wildlife. The situation is different in farmed deer: a herpesvirus specific of red deer (Cervus elaphus) was isolated in Scotland (Inglis at al., 1983) and in Denmark (Ronsholt et al., 1987) and episodes of malignant catarrhal fever have also been described (Reid et al., 1979).

After infection, herpesviruses persist in their infected host in a latent state. A lifelong latent infection is established. Therefore, latency allows the persistence of a herpesvirus in a population of restricted size. Several stimuli are able to induce reactivation of latent herpesvirus and its subsequent reexcretion. Usually, reactivated animals are not ill, but they can transmit the virus to their surroundings (Thiry et al., 1986). When reactivation occurs in wild animals, the risk of transmission of the virus is very low. It increases when animals are gathered, for example in a deer farm. Therefore, even a low prevalence of herpesvirus infection must be carefully considered, because it indicates that the herpesvirus is present and may be maintained in the population. A clinical disease provoked by this virus may be observed in gathered animals.

The prevalence of herpesvirus infections in wild ruminants is not well documented in continental Europe. Red deer and roe

deer (_Capreolus__capreolus_) seropositive against infectious
bovine rhinotracheitis virus (bovine herpesvirus 1; BHV-1) have
been observed (Kokles et al., 1977; Lawman et al., 1978). The
recent isolation of two herpesviruses specific to red deer
(Inglis et al., 1983, Ronsholt et al., 1987) and to reindeer
(_Rangifer_tarandus_) (Ek-Kommonen et al., 1986) has modified the
understanding of results of previous serological surveys,
because these viruses are serologically related to BHV-1 (Thiry
et al., 1988a).

Herpesvirus infections of red deer and other wild
ruminants are reviewed: infections with BHV-1 and other
serologically related herpesviruses; infections with bovine
herpesvirus 2 (BHV-2) and bovid herpesvirus 4 (BHV-4) and with
herpesviruses serologically related to alcelaphinae herpesvirus
1 (AHV-1).

THE RED DEER IN BELGIUM

The size of the red deer population in Belgium has
fluctuated since the nineteenth century. In 1850, only one
small population still existed: 3 to 4 red deer were still
living in Freyr Forest (De Crombrugghe, 1978; Libois, 1983).
Thereafter, the size of the red deer population increased by
the introduction of captive red deer and animals imported from
Eastern Europe. In 1880, there were about 2000 red deer. The
population size decreased during the two world wars and, in
1980, 4000 to 6000 red deer were living in Belgium. There is
actually no deer farm in Belgium. Fallow deer (_Dama_dama_) is
usually bred in closed areas. It is free-living only in
Ciergnon forest (50 to 100 individuals) (Libois, 1983).
Therefore, the actual red deer population is originated from a
very small number of animals, but also from introduced animals.
Red deer is actually living in forests in the south of Belgium
(Figure 1).

Fig. 1   Geographical distribution of red deer in Belgium
(de Crombrugghe, 1978).

INFECTIONS WITH HERPESVIRUSES SEROLOGICALLY RELATED TO BOVINE HERPESVIRUS 1

Four herpesviruses isolated from ruminants are serologically related to BHV-1: caprine herpesvirus 2 (CHV-2) was isolated from domestic goat (Mettler et al. 1979); CHV-2 infection is associated with enteritis in young goat, abortion and vulvovaginitis; herpesvirus of Cervidae type 1 (HVC-1) was isolated from red deer showing ocular disease (Inglis et al., 1983; Ronsholt et al., 1987); reindeer herpesvirus was isolated from vaginal swabs after dexamethasone treatment (Ek-Kommonen et al., 1986) and buffalo herpesvirus was isolated from water buffalo (Bubalus arnee) (St George and Philpott, 1972; Brake and Studdert, 1985). A ruminant species seropositive for BHV-1 may be therefore infected by BHV-1 or one of these related viruses. Antibodies against BHV-1 were previously demonstrated in roe deer and red deer, in Great Britain (Lawman et al., 1978) and East Germany (Kokles, 1977). No roe deer seropositive against BHV-1 was observed in Trois Fontaine forest in France (Blancou, 1983). Antibodies against BHV-1 and CHV-2 were not detected in wild ruminants in Switzerland (Hasler and Engels, 1986). The results of a serological survey carried out between 1981 and 1986 in France and Belgium are given in Tables 1 to 3 (Thiry et al., 1988b). Antibodies against each of the four viruses were titrated by seroneutralisation. Titres of virus suspensions used in this test were estimated by plaque assay. Physical virus particles were not titrated so that antibody titres against each virus cannot be properly compared.

TABLE 1   Serological results of red deer for six herpesviruses in France and Belgium, 1981-1986. The percentage of seropositive (titre >=8) animals is given into brackets; Total: number of animals seropositive against at least one of the 4 viruses: BHV-1, CHV-2, HVC-1 and reindeer herpesvirus.

YEAR	N	BHV-1	CHV-2	HVC-1	Reindeer herpesvirus	Total	BHV-2	BHV-4
FRANCE								
1982	48	0	2	2	2	2	0	0
1983	22	0	0	0	0	0	0	0
1984	5	0	0	0	0	0	0	0
1985	5	0	0	0	0	0	0	0
TOTAL	80	0	1	1	1	1	0	0
BELGIUM								
1985	33	9	6	6	9	12	0	0
1986	37	3	9	1	3	11	0	0
TOTAL	70	6	7	8	6	11	0	0

Only one red deer (1%) was positive in France against three of these serologically related herpesviruses. In contrast, 8 out of the 70 red deer (11%) sampled in Belgium in 1985 and 1986 were seropositive against at least one of the

four related herpesviruses (Table 1). No antibodies against
these virus were detected in sera of 80 roe deer sampled in
Belgium in 1981, 1985 and 1986. These results were similar to
these obtained with 387 roe deer sera sampled in France between
1982 and 1986 (<1% seropositive) (Table 3). No mouflon (ovis
ammon musimon) were seropositive, but the number of mouflon
investigated was very low. Only one serum out of 28 ibex (Capra
ibex) (4%) from France was positive against CHV-2. Four out of
99 chamois (4%) (Rupicapra rupicapra) were positive to one of
the four related herpesviruses (Table 4) (Thiry et al., 1988b).

TABLE 2   Results of serological tests for viruses
serologically related to BHV-1 in red deer from Belgium;
titres are expressed as the reciprocal of the highest
dilution of serum inhibiting 50% of cytopathic effect.

	BHV-1	CHV-2	HVC-1	Reindeer herpesvirus
1985				
BC6 CIERGNON F. ADULT		64		
BC11 CIERGNON F. ADULT	32	8	64	32
BC29 ANLIER M. ADULT	16			16
BC38 CIERGNON F. ADULT	8		16	16
1986				
BC3 ELSENBORN –	16	64	64	64
BC46 CIERGNON F. ADULT			16	
BC60 GRAND-HALLEUX M. YOUNG			16	
BC73 SAINT-HUBERT F. ADULT		64	64	

Several ruminant species appear therefore to be infected
with a herpesvirus serologically related to BHV-1. The
experimental inoculation of red deer with BHV-1 produces
asymptomatic infection (Reed et al., 1986). It is therefore
assumed that red deer seropositive for one of the related
viruses have been infected by HVC-1. The geographical
distribution of seropositive red deer in Belgium is given in
Figure 2. Four out of the 8 positive red deer were living in
Ciergnon forest. The other 4 positive cases were observed in
other forests (Table 4).

INFECTION WITH BOVINE HERPESVIRUS 2

BHV-2 infection provokes in Europe infectious bovine
mammillitis. Lesions are usually localized on the skin of the
udder. In Africa, a generalized skin disease is observed and is
called pseudo-lumpy skin disease.
The prevalence of cattle seropositive to bovine
herpesvirus 2 (BHV-2) was 28% in Belgium (Pastoret et al.,
1983). This infection is therefore wide-spread in this country.
Nevertheless, no wild ruminants (red deer or roe deer)
wereseropositive for BHV-2 between 1981 to 1986 in Belgium
(Tables 1 and 3) (Thiry et al., 1988b). In France, only a low
prevalence of wild ruminant positive for BHV-2 was

demonstrated: < 1% in roe deer and 1% in chamois (Tables 3 and 4). The epizootiology of BHV-2 infection is different in Africa, where this infection is prevalent in more than 20 species of wild ruminants (Plowright and Jesset, 1971; Hamblin and Hedger, 1982). This infection is at least very rare in wild ruminants from France and Belgium and no diseasesassociated with BHV-2 have been so far recognized.

TABLE 3   Serological results of roe deer for six herpesviruses in France and Belgium, 1981-1986. See table 1 for explanations.

YEAR	N	BHV-1	CHV-2	HVC-1	Reindeer herpesvirus	Total	BHV-2	BHV-4
FRANCE								
1982	7	O	O	O	O	O	O	O
1983	59	O	O	O	O	O.	3.5	O
1984	52	O	2	O	O	2	2	O
1985	69	3	1.5	3	1.5	3	O	O
1986	200	O	O	O	O	O	O	O
TOTAL	387	<1	<1	<1	<1	<1	<1	O
BELGIUM								
1981	21	O	O	O	O	O	O	O
1985	20	O	O	O	O	O	O	O
1986	39	O	O	O	O	O	O	O
TOTAL	80	O	O	O	O	O	O	O

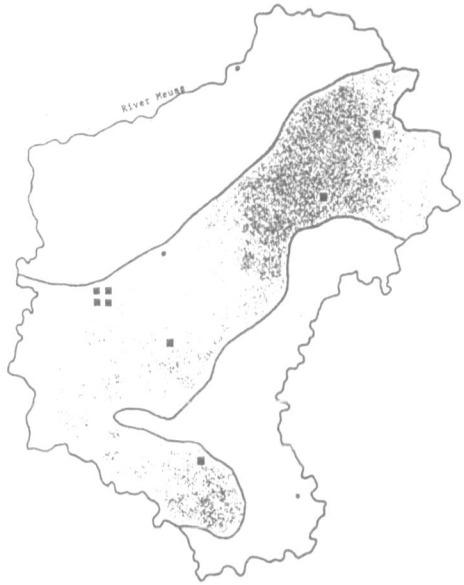

Fig. 2   Geographical distribution of red deer sero-positive for herpesviruses serologically related to BHV-1 in Belgium; ● positive animals.

INFECTION WITH BOVID HERPESVIRUS 4

Most of the BHV-4 infections of cattle are sub-clinical. This virus has been associated with a variety of symptoms: respiratory, skin or genital diseases. It was recently demonstrated that this virus provokes post-partum metritis in cattle (Wellemans et al., 1986).

The prevalence of BHV-4 infection in cattle is 22,5 % in Belgium (Van Malderen et al., 1987). This infection has been identified in cattle in several countries throughout the world: Africa, U.S.A., Italy, Switzerland, West Germany and Belgium. A serological survey was undertaken in wild ruminants from France and Belgium between 1981 and 1986.

An indirect fluorescent antibody test was used in this experiment. No red deer or other wild ruminant from France and Belgium was seropositive for BHV-4 (Tables 1, 3 and 4) (Thiry et al., 1988b).

TABLE 4. Serological results of French Chamois-Isards, ibex and mouflons for six herpesviruses, 1982-1986. See table 1 for explanations.

YEAR	N	BHV-1	CHV-2	HVC-1	Reindeer herpesvirus	Total	BHV-2	BHV-4
CHAMOIS								
1982	15	O	O	6.6	O	6.6	6.6	O
1983	51	O	O	4	O	4	O	O
1984	13	8	8	8	8	8	O	O
1985	3	O	O	O	O	O	O	O
1986	17	O	O	O	O	O	O	O
TOTAL	99	1	1	4	1	4	1	O
IBEX								
1984	7	O	O	O	O	O	O	O
1985	3	O	O	O	O	O	O	O
1986	18	O	5	O	O	5	O	O
TOTAL	28	O	4	O	O	4	O	O
MOUFLON								
1986	4	O	O	O	O	O	O	O

INFECTION WITH HERPESVIRUSES SEROLOGICALLY RELATED TO ALCELAPHINE HERPESVIRUS 1

Alcelaphine herpesvirus 1 (AHV-1) provokes malignant catarrhal fever (MCF) in African cattle. MCF is a generalised disease which occurs in cattle following transmission of AHV-1 from the natural host, the wildebeest (Connochaetes sp.) (Reid and Buxton, 1985). The European form of MCF occurs in cattle and deer following contact with sheep (Reid et al., 1985) which allows the transmission of the agent. This "sheep associated agent" has not yet been identified, but it is probably serologically related to AHV-1. Sheep possess indeed antibodies against a virus related to AHV-1 (Rossiter, 1981).

Serological studies carried out in zoological gardens have indeed shown a significant prevalence (>10%) of antibodies against herpesviruses related to AHV-1 in chamois, mouflon and ibex (Plowright, 1986). It is not yet known if wild Caprinae are also seropositive for herpesviruses related to AHV-1. European form of MCF is fatal in deer. The presence of antibodies in these species is therefore less probable.

DISCUSSION

The prevalence of herpesvirus infections in wild ruminants is very low in France and Belgium. A striking feature is the difference observed between infections of red deer with viruses related to BHV-1 in France and in Belgium. The latent infection with such herpesviruses could be present in the Belgian red deer population since a long time. As this population arose from a small number of animals, this infection could be maintained or even introduced by a red deer imported from another country. This situation allowed the circulation of the latent virus in a population of restricted size. This observation supports the hypothesis that HVC-1 is present in Belgium, but no clinical ocular disease has been until now observed. The prevalence of HVC-1 infection is higher in Great Britain: 33 to 40% of seropositive Scottish red deer and 14% of seropositive English deer compared to 11% of seropositive red deer in Belgium (Nettleton et al., 1986; Thiry et al., 1988b).

Deer are not in direct contact with cattle. Such contacts are usually required to transmit efficiently herpesviruses. This explains why infections with herpesviruses which are highly prevalent in cattle (BHV-2, BHV-4) are not significant in wild ruminants. Therefore wild species are probably infected by their own herpesvirus: HVC-1 in red deer and reindeer herpesvirus in reindeer. Alternatively, they may be infected by a herpesvirus which naturally infects a closely related species. For example, chamois seropositive to a herpesvirus related to BHV-1 may be infected by CHV-2, but this has not yet been demonstrated. Nevertheless herpesvirus infection of a species different from the natural host is often asymptomatic: for example, BHV-1 infection of domestic goat (Pirak et al., 1983) or of red deer (Reid et al., 1986). A further work will consist in the isolation and the characterization of herpesviruses carried by wild ruminants in France and Belgium.

ACKNOWLEDGEMENTS

We wish to thank M. Muys for typing the manuscript. We thank also P.F. Nettleton (Edinburgh), C. Ek-Kommonen (Helsinki), H. Ludwig (Berlin), G. Castrucci (Perugia) for providing us with virus strains.

156

REFERENCES

Blancou, J. 1983. Serologic testing of wild roe deer (Capreolus capreolus L.) from the Trois Fontaines forest region of Eastern France. J. Wildl. Dis., 19, 271-273.

Brake, F., Studdert, M.J. 1985. Molecular epidemiology and pathogenesis of ruminant herpesviruses including bovine, buffalo and caprine herpesvirus 1 and bovine encephalitis herpesvirus. Aust. Vet. J., 62, 331-334.

Ek-Kommonen, C., Pelkonen, S., Nettleton, P.F. 1986. Isolation of a herpesvirus serologically related to bovine herpesvirus 1 from a reindeer (Rangifer tarandus). Acta Vet. Scand., 27, 299-301.

De Crombrugghe, S. 1978. Dynamique des populations et gestion des grands mammifères dans l'Hertogenwald. In: Journées d'Etudes sur les Problèmes liés à l'Etude et à la gestion des Hautes Fagnes et de la Haute Ardenne, Liège, Belgique, 167-195.

Hamblin, C., Hedger, R.S. 1982. Prevalence of neutralizing antibodies to bovid herpesvirus 2 in African wildlife. J. Wildl. Dis., 18, 429-436.

Hasler, J., Engels, M. 1986. Stellen nichtbovine Paarhufer ein IBR-virus-Reservoir dar? II. Seroepidemiologische Unter-suchungen an Ziegen, Schafen, Schweinen und Wild-paarhufern in der Schweiz. Schweiz. Arch. Tierheilk., 128, 575-585.

Inglis, D.M., Bowie, J.M., Allan, M.J., Nettleton, P.F. 1983. Ocular disease in red deer calves associated with a herpesvirus infection. Vet. Rec., 113, 182-183.

Kokles, R. 1977. Untersuchungen zum Nachweis von IBR/IPV-Antikörpen bei verschiedenen Haus-und Wildtieren sowie beim Menschen. Monat. Vet. Med., 32, 170-171.

Lawman, M.J.P., Evans, D., Gibbs, E.P.J., Mc Diarmid, A., Rowe, L. 1978. A preliminary survey of British deer for antibody to some virus diseases of farm animals. Br. Vet. J., 134, 85-91.

Libois, R. 1983. Le Cerf. In: Animaux menacés de Wallonie. Protégeons nos mammifères. Duculot-Région Wallonne, Gembloux, Belgique, p.133-144.

Mettler, F., Engels, M., Wild, P., Bivetti, A. 1979. Herpesvirus-infektion bei Zincklein in der Schweiz. Schweiz. Arch. Tierheilk., 121, 655-661.

Nettleton, P.F., Sinclair, J.A., Herring, J.A., Inglis, D.M., Fletcher T.J., Ross, H.M., Bonniwell, M.A. 1986. Prevalence of herpesvirus infection in British red deer and investigations of further disease outbreaks. Vet. Rec., 118, 267-270.

Pastoret, P.-P., Antoine, H., Schwers, A., Thiry, E., Castrucci, G. 1983. Enquête sérologique sur l'infection par le virus de la mammillite herpétique bovine (Bovine herpesvirus 2, BHV-2) en Belgique. Ann. Méd. Vét., 127, 267-270.

Pirak, M., Thiry, E., Brochier, B., Pastoret P.-P. 1983. Infection expérimentale de la chèvre par le virus de la rhinotrachéite infectieuse bovine (Bovine herpesvirus 1) et tentative de réactivation virale. Rec. Méd. Vét., 159, 1103-1106.

Plowright, W., Jessett, D.M. 1971. Investigations of Allerton-type herpesvirus infection in East African game animals and cattle. J. Hyg., 69, 209-222.

Plowright, W. 1986. Malignant catarrhal fever. Rev. Sci. Techn. Off. int. Epiz., 5, 897-918.

Reid H.W., Buxton, D., Corrigall, W., Hunter, A.R., Mc Martin, D.A., Rushton, B. 1979. An outbreak of malignant catarrhal fever in red deer (Cervus elaphus). Vet. Rec., 104, 120-123.

Reid, H.W., Buxton, D. 1985. Immunity and pathogenesis of malignant catarrhal fever. In: (Ed.: P.-P. Pastoret, E. Thiry, J. Saliki) Immunity to herpesvirus infections of domestic animals. Commission of the European Communities, EUR 9737, Luxembourg, p.117-130.

Reid, H.W., Nettleton, P.F., Pow, I., Sinclair, J.A. 1986. Experimental infection of red deer (Cervus elaphus) and cattle with a herpesvirus isolated from red deer. Vet. Rec., 1986, 118, 156-158.

Ronsholt, L., Siig Christensen, L., Bitsch, V. 1987. Latent herpesvirus infection in red deer: characterization of a specific deer herpesvirus including comparison of genomic restriction fragment pattern. Acta Vet. Scand., 28, 23-31.

Rossiter, P.B. 1981. Antibodies to malignant catarrhal fever virus in sheep sera. J. Comp. Pathol., 91, 303-311.

St George, T.D., Philpott, M. 1972. Isolation of infectious bovine rhinotracheitis virus from the prepuce of water buffalo bulls in Australia. Aust. Vet. J., 48, 126.

Thiry, E., Dubuisson, J., Pastoret, P.-P. 1986. Pathogenesis, latency and reactivation of infections by herpesviruses. Rev. Sci. techn. Off. int. Epiz., 5, 809-819.

Thiry, E., Meersschaert, C., Pastoret, P.-P. 1988a. Epizootiologie des infections à herpèsvirus chez les ruminants sauvages. I. Le virus de la rhinotrachéite infectieuse bovine et les virus antigéniquement apparentés. Rev. Elev. Méd. Vét. Pays Trop., submitted for publication.

Thiry, E., Vercouter, M., Dubuisson, J., Barrat, J., Sepulchre, C., Gerardy, C., Meersschaert, C., Collin, B., Blancou, J., Pastoret, P.-P. 1988b. Serological survey of herpesvirus infections in wild ruminants of France and Belgium. J. Wildl. Dis., in press.

Van Malderen, G., Van Opdenbosch, E., Wellemans, G. 1987. Bovine herpesvirus 1 and 4: a sero-epidemiological survey of the Belgian cattle population. Vlaams Diergeneesk. Tijdschr., 56, 364-371.

Wellemans, G., Van Opdenbosch, E., Mammerickx, M. 1986. Inoculation expérimentale du virus LVR 140 (herpès bovin IV) à des vaches gestantes et non gestantes. Ann. rech. Vét., 17, 89-94.

THE DIAGNOSIS OF MALIGNANT CATARRHAL FEVER IN DEER

D. Buxton

Moredun Research Institute, Edinburgh EH17 7JH.

ABSTRACT

Farmed deer are very susceptible to MCF, a fatal disease thought to be caused by a virus transmitted from normal sheep. An almost identical condition can occur in susceptible animals where the cause is alcelaphine herpesvirus 1, transmitted from wildebeest. Both causal agents are related but as the putative sheep agent has not yet been identified MCF of farmed deer remains a clinicopathological entity. Diagnosis is achieved by assessing clinical, post mortem and histopathological findings. Virus isolation and serology are not feasible means of diagnosis.

MCF has been recorded in several species of deer and can present as peracute (sudden death), acute, subacute and chronic forms. Typically animals first appear dull and febrile. Nasal and ocular discharges, initially serous, become muco-purulent and conjunctivitis and corneal opacity develop. In the buccal cavity haemorrhagic erosions occur while surface lymph nodes are enlarged and readily palpated. There may be scabby lesions on the skin, in the perineal region, over the shoulders and around the mouth. Diarrhoea or dysentery is frequently present and its severity can influence the duration of illness.

At post mortem examination additional lesions can include haemorrhagic intestinal mucosa and gut contents and focal haemorrhages in the urinary bladder. The kidneys are characteristically covered in small raised white foci and the liver may be swollen and pale. The enlarged lymph nodes may appear pink soft and necrotic.

Diagnosis however can only be confirmed by the identification of typical histological lesions which generally consist of epithelial degeneration with associated lymphoid infiltration and capillary haemorrhages, vasculitis, hyperplasia and necrosis of lymphoid organs and interstitial infiltrations and accumulations of lymphoid cells in non-lymphoid tissues.

INTRODUCTION

Farmed deer are very susceptible to MCF, a fatal disease of cattle, and deer (Plowright, 1986). The cause of MCF in deer is not known but it is thought to be due to a virus carried by clinically normal sheep. In Europe MCF can also occur in susceptible zoo animals due to alcelaphine herpesvirus 1, transmitted from clinically normal wildebeest.

Epizootics of MCF have been reported in farmed deer from the UK, Australia and New Zealand where it is recognised as the most serious

159

infectious disease affecting the industry. Following the initial outbreak in which 9/15 red deer (<u>Cervus elaphus</u>) died (Reid <u>et al</u>., 1979) there have been several reports in which the herd mortality was in excess of 50%. Sika deer (<u>Cervus nippon</u>) (Clark <u>et al</u>., 1971), rusa deer (<u>Cervus timorensis</u>) (Denholm and Westbury, 1982) and Père David's deer (<u>Elaphurus davidianus</u>) have all been affected also. In recent attempts to develop this latter species for commercial venison production in New Zealand and the UK very many have died from MCF (Reid <u>et al</u>., 1987; Orr and Mackintosh, 1987) and it has been recommended that no further attempts to exploit this species should be made until the disease is more fully understood. Species of deer not currently farmed are also very susceptible to MCF.

As MCF is a clinicopathological entity, diagnosis is achieved by assessing clinical, post mortem and histopathological findings and virus isolation or serology are not feasible at present. The description below was taken from the following reports of MCF in deer: Huck <u>et al</u>., 1961; Clark <u>et al</u>., 1971; Wyand <u>et al</u>., 1971; Reid <u>et al</u>., 1979; Denholm and Westbury, 1982; Oliver <u>et al</u>., 1983; Wilson <u>et al</u>., 1983; Orr and Mackintosh, 1987; Reid <u>et al</u>., 1987; unpublished data of the author.

CLINICAL SIGNS

MCF can present as peracute, acute, subacute or chronic disease, the clinical signs becoming progressively more marked with the duration of illness. In peracute cases the animal may die with no prior clinical signs having been noticed and in these animals a diagnosis will depend on a careful post mortem examination and the histology of a wide selection of tissues.

In other cases the animal is first noticed as being dull and off its food. The rectal temperature is initially elevated to 41 or 42°C and the pulse rate is also increased, although both can return to normal in chronically affected animals. At the same time salivation and a clear watery nasal discharge is seen together with extensive bilateral lacrimation, initially limited to the inner canthus. If the animal survives longer the discharges become muco-purulent

With the onset of a lacrimal discharge bilateral congestion of the conjunctiva and sclera occurs together with a progressive opacity of the cornea which starts at the limbus and proceeds centripetally until, in chronic cases, the whole cornea is clouded. In the peracute disease there may be no discernible opacity, although mild microscopic lesions will usually be present, while in rare cases of chronic MCF the cornea can become eroded and ulcerated. Photophobia may also occur, with swelling of the eyelids and catarrhal matting of the eyelashes, so that the eyes become closed.

In chronic cases the muzzle may be initially dry and hot, with hyperaemia of the lining of the nares, but as the discharge becomes catarrhal the nares become blocked with sticky encrusting secretion causing snuffling sounds. In severe cases blockage of the nares necessitates mouth breathing. The epidermis of the muzzle can become cracked in the few animals that survive for several days, becoming progressively more severe with eventual extensive sloughing and bleeding.

With the initial rise in temperature there is excessive saliva inside the mouth, the oral mucosa is hyperaemic and shallow focal erosions may become visible over the next 24 to 48 hours. They are found most readily on the tips of the buccal papillae and to a lesser extent at the commissures of the lips, on the dorsum of the tongue and on the hard palate. In peracute cases macroscopic lesions may not be seen. In hinds the vulval mucous membranes may become reddened before the appearance of foci of yellowish encrustation.

Deer often present with acute diarrhoea or dysentery from the onset of illness and their urine may be dark and contain blood.

Lymph nodes are usually enlarged in MCF and hence those located superficially such as the submandibular, prescapular and prefemoral nodes, are readily palpated. Scabby lesions may also occur in the skin in the perineal region.

Nervous signs are not uncommon with initial dullness proceeding to hyperaesthesia and inco-ordination. The animals joints are sometimes "puffy" and swollen.

Haematological investigations indicate that an initial rise in circulating white blood cells is followed by a marked leucopaenia. At the same time the proportion of neutrophils may rise and show a "shift to the left". The red blood cell count, packed cell volume and haemoglobin estimations increase in cases of diarrhoea due to dehydration and resultant haemoconcentration.

MACROSCOPIC PATHOLOGICAL FINDINGS

In peracute cases very few gross changes may be observed at post mortem exmination. However in other cases in addition to the lesions seen clinically, other macroscopic changes are widespread. Within the buccal cavity and pharynx, and including the soft palate and the tongue, lesions can be found varying from distinct red erosions and ulcers to more diffuse patches of necrosis. Similar but generally milder lesions are sometimes also found in the oesophagus and forestomachs. The abomasal mucosa can appear reddened, sometimes with haemorrhagic stripes along the mucosal folds.

In the intestines, congestion and oedema of the wall may extend from the duodenum to the rectum, there may be focal haemorrhages in the mucosa and the intestinal contents will often contain blood and be watery at all levels. The mesenteric lymph nodes in deer are characteristically grossly enlarged and can be surrounded by translucent yellowish oedema. While usually firm and white on cross section they may sometimes be necrotic and haemorrhagic. Other carcase lymph nodes are similarly enlarged with the retropharyngeal nodes apparently more commonly appearing necrotic and haemorrhagic. The spleen may be slightly shrunken. The liver is either normal or swollen and congested with pale areas visible on its surface.

Lesions in the respiratory system may range from congestion of the mucosa covering the nasal septum and turbinates, larynx, trachea and bronchi to extensive ulceration and diphtheritic deposits and haemorrhages. Foci of cellular consolidation may be scattered throughout the parenchyma of the lung.

One of the most characteristic lesions occurs in the kidney where, under the capsule, there are raised white foci from one to four mm. in diameter, sometimes surrounded by a thin haemorrhagic zone. Examination of the cut surface shows that they are scattered through the cortex. The epithelial surface of the urinary bladder is commonly covered in haemorrhagic foci of irregular size and shape.

Genital tract lesions are rare and generally confined to superficial erosions of the vaginal mucosa. The joints can contain excess fluid and the synovial membranes may appear swollen and reddened, while the brain may be congested and bathed in excess and slightly cloudy CSF.

MICROSCOPIC PATHOLOGICAL FINDINGS

These can generally be divided into epithelial degeneration, vasculitis, lymphoid hyperplasia and interstitial infiltrations and accumulations of lymphoid cells in non lymphoid tissues.

Epithelial lesions are essentially similar whether they occur in the buccal or nasal cavities, lungs, alimentary tract, gall bladder, urinary bladder, skin, or conjunctiva. They are also associated with subepithelial and intraepithelial lymphoid cell infiltrates and sometimes also with vasculitis and haemorrhages. With stratified squamous epithelia, acantholysis and ortho- and para-keratotic hyperkeratosis occur and this can give way to erosions and ulcerations (Figure 1). In the case of respiratory and intestinal mucosae there can be cellular degeneration and sloughing and also accretions of fibrinonecrotic exudate on the surface of the lesion.

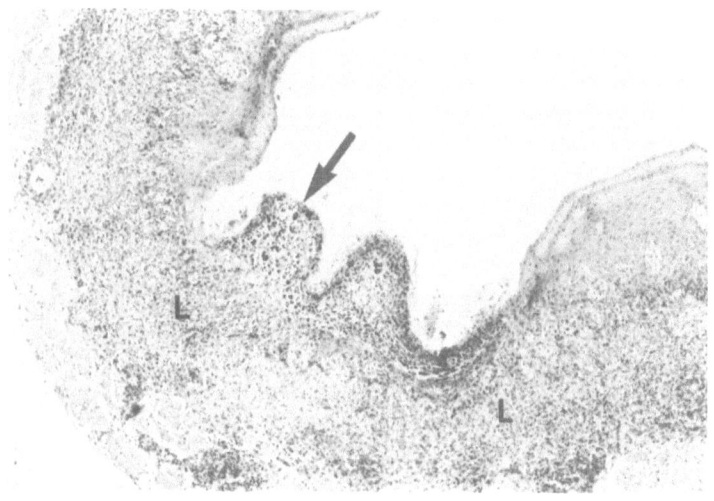

Fig. 1.   Oesophagus of a red deer (Cervus elaphus) with MCF.   Note the disrupted stratified squamous epithelium (arrow) infiltrated by lymphoid cells.   A similar infiltrate (L) is present in the lamina propria and submucosa.   HE x 54.

Lymph node hyperplasia arises as a result of a marked expansion of lymphoblastoid cells, which are often seen to be undergoing mitoses, in the paracortex. The cortex is also hyperplastic although to a lesser extent, and there is generally little follicular development although there may be exceptions. When necrosis is present it appears to be follicular in origin although in advanced cases it can involve most structures in the node. Haemorrhages also occur, perhaps as a result of vasculitis (see below). Medullary cords are thickened and the sinuses packed with macrophages and lymphoid cells. Periglandular oedema and lymphoid inflammation are also common.

In the spleen the tissue is substantially depleted of lymphoid cells and only small 'islands' of lymphoid cells represent the periarteriolar lymphoid sheaths.

Interstitial infiltrations and accumulations of lymphoid cells in non lymphoid tissues such as the periportal areas of the liver and the interstitium of the renal cortext (Figure 2) are characteristic of MCF.

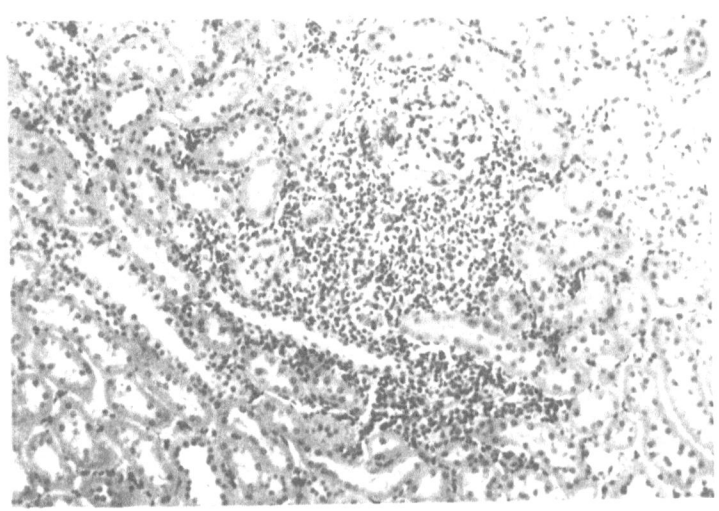

Fig. 2. Interstitial lymphoid cells in the kidney of a red deer (Cervus elaphus) with MCF. HE x 136

In the latter, accumulations are often so large that they are readily appreciated as macroscopically visible white foci on the surface of the kidney (see above). Other commonly affected tissues include salivary and lacrimal glands, pancreas and cardiac and skeletal muscle. It is worth noting that generally the lymphoid accumulations are not associated with necrosis of adjacent tissues, muscle being the exception.

In the brain there is often a non-suppurative meningoencephalitis with perivascular cuffing by lymphoid cells. Some associated small foci of microglial proliferation and periaxonal oedema can occur. The choroid plexus can be infiltrated by lymphoid cells and the cerebrospinal fluid contains unusually large numbers of mononuclear cells.

In the eye the principal lesion is a lymphoid, interstitial keratitis originating at the limbus and progressing centrally. Neutrophil infiltrates occur when lesions have advanced to corneal ulceration.

Vasculitis is invariably a feature of MCF and can affect arteries (Figure 3), arterioles, veins and venules. Lymphoid cells are found in the tunica adventitia and tunica media, in which there is often

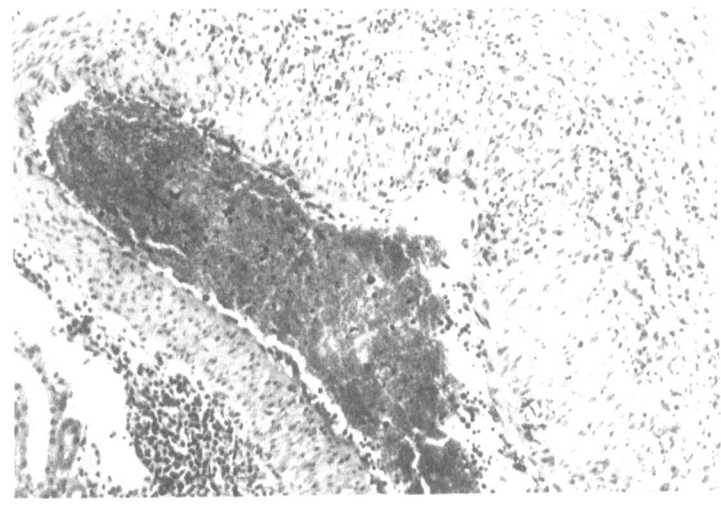

Fig. 3. Arteritis in the medulla of the kidney of a red deer (<u>Cervus elaphus</u>) with MCF. HE x 136

fibrinoid degeneration. The intima is also involved with endothelial cell hypertrophy and degeneration, pavementing of lymphoid cells on the luminal surface, as well as subendothelial inflammatory cell accumulations. In severe cases occlusion of the lumen by lymphoid cells and hyperplastic enothelial cells can sometimes occur. Haemorrhages are often associated with affected vessels.

DIFFERENTIAL DIAGNOSIS

The differential diagnosis of MCF in deer should seek to exclude other causes of sudden death such as clostridial disease and anthrax. Yersiniosis, a condition associated with stress may also present acute symptoms similar to MCF although ocular and oral lesions will not be present and surface lymph nodes do not normally show gross enlargement. Nervous symptoms seen in cases of louping-ill virus infection or salmonellosis should also be distingusihed from cases of MCF. In salmonellosis diarrhoea or dysentery may also be encountered. However it must be stressed that in some peracute cases of MCF, lesions may only be visible upon histological examination. In all cases of sudden death in deer where clinical and macroscopic changes do not allow a clear diagnosis a wide range of tissues should be examined for histological evidence of MCF.

REFERENCES

Clark, K.A., Robinson, R.M., Marburger, R.G., Jones, L.P. and Orchard, J.H. (1971). Malignant catarrhal fever in Texas cervids. J. Wildlife Dis., 6, 376-383.

Denholm, L.J. and Westbury, H.A. (1982). Malignant catarrhal fever in farmed rusa deer (Cervus timorensis). 1. Clinico-pathological observations. Aust. Vet. J., 58, 81-87.

Huck,R.A., Shand, A., Allsop, P.J. and Paterson, A.B. (1961). Malignant catarrh of deer. Vet. Rec., 73, 457-465.

Oliver, R.E., Beatson, N.S., Cathcart, A. and Poole, W.S. (1983). Experimental transmission of malignant catarrhal fever to red deer (Cervus elaphus). N.Z. Vet. J., 31, 209-212.

Orr, M. and Mackintosh, C. (1987). Outbreak of malignant catarrhal fever in Pere David's deer. Proceedings of a deer course for veterinarians. Deer branch of the New Zealand Veterinary Association. Number 4, 181-185.

Plowright, W. (1986). Malignant catarrhal fever. Rev. Sci. Tech. Off. Int. Epiz. 5, 939-958.

Reid, H.W., Buxton, D., Corrigall, W., Hunter, A.R., McMartin, D.A. and Rushton, R. (1979). An outbreak of malignant catarrhal fever in red deer (_Cervus elaphus_). Vet. Rec., 104, 120-123.

Reid, H.W., Buxton, D., McKelvey, W.A.C., Milne, J.A. and Appleyard, W.T. (1987). Malignant catarrhal fever in Pere David's deer. Vet. Rec. 121, 276-277.

Wilson, P.R., Alley, M.R. and Irving, A.C. (1983). Chronic malignant catarrhal fever: a case in a sika deer (_Cervus nippon_). N.Z. Vet. J., 31, 7-9.

Wyand, D.S., Helmboldt, C.F. and Nielsen, S.W. (1971). Malignant catarrhal fever in white-tailed deer. J.A.V.M.A., 159, 605-610.

THE AETIOLOGY OF MALIGNANT CATARRHAL FEVER

H.W. Reid

Moredun Research Institute, 408 Gilmerton Road,
Edinburgh, EH17 7JH, Scotland

ABSTRACT

Despite the dramatic pathological changes associated with Malignant Catarrhal Fever (MCF) and considerable research effort worldwide, the aetiology remains obscure. With the identification of this condition as the most frequently encountered infectious disease problem in farmed deer elucidation of the nature of the causal agent is essential to the industry. It is known that a herpesvirus which normally infects wildebeest subclinically can spread to other ruminants and cause MCF. The epidemiology of MCF where wildebeest are not present suggests that sheep act as reservoirs of infection but no virus has been isolated.

Studies to identify this putative virus of sheep using reagents generated from the wildebeest virus will be described in this paper. The evidence suggests that deer and cattle become infected from sheep with a virus closely related antigenically and genomically, to but distinct from, the wildebeest virus. This virus spreads very efficiently amongst domestic sheep, and probably all are infected at a young age. Deer which are much more susceptible to infection than cattle become infected with the virus and develop MCF. However virus expression would appear to be incomplete in affected animals explaining why no agent can be recovered. Investigations that will clarify these suggestions will be described.

INTRODUCTION

It is extraordinary that the cause of Malignant Catarrhal Fever (MCF) which is one of the most dramatic virus-induced animal diseases is still poorly understood. That MCF has emerged as the single most important disease of farmed deer (Anon 1980) gives an urgency to resolving this problem if the industry is to prosper.

MCF was originally reported in Europe (for review see Plowright, 1986) as a disease of cattle but has subsequently been recognised to occur worldwide (Odend'Hal, 1983). Two forms of the disease which are clinically and pathologically indistinguishable are recognised: one prevalent in Africa where wildebeest (Connochaetes sp.) are present (Mettam 1924) and one that occurs following contact with sheep (Gotze 1932). They are referred to as wildebeest associated (WA) and sheep associated (SA) MCF respectively. A gammaherpesvirus designated

Alcelaphine herpesvirus-1 (AHV-1) (Reid, et al., 1975) which can induce MCF in cattle and some other ruminants has been isolated from wildebeest (Plowright et al., 1960) but no virus has been identified in the SA form of the disease. This putative virus of sheep known as the SA-agent (SA-A) is the cause of MCF in farmed deer in Europe, Australia and New Zealand (Reid et al., 1979, Denholm and Westbury 1982, McCallum et al.,1982).

In this paper current understanding of the viruses that cause MCF will be briefly reviewed and recent developments with regard to identifying the SA-A described.

HOST RANGE

It is probable that all wildebeest are infected with AHV-1 either in utero or during the first few months of life (Plowright 1965, 1967). Infection in wildebeest is however subclinical and it is only when the virus is transmitted to other members of the families Bovidae and Cervidae that the reaction known as MCF occurs (Plowright 1986). Laboratory rabbits and rodents also react similarly to experimental infection (Daubney and Hudson 1936, Jacoby et al., 1988).

Examination of sera from other large antelope species of the subfamilies Alcelaphinae and Hippotraginae, for neutralising antibody to AHV-1, indicates that similar viruses are prevalent as inapparent infections in these species (Reid et al., 1975; Hamblin and Hedger, 1984). A conclusion supported by the isolation of viruses from Alcelaphus buselaphus, A. caama, Damaliscus korrigum, D. lunatus, Connochaetes gnu and Oryx dammah which are probably distinct from but closely related to AHV-1 (Reid et al., 1975; Mushi et al., 1981; Heuschele et al., 1984). Furthermore antibody to AHV-1 has also been detected in sera from 12 species of the subfamily Caprinae (Heuschelle et al., 1984) and using the less discriminating indirect immunofluorescent (IIF) test, antibody has also been detected in domestic sheep and goats (Rossiter 1981, Reid, unpublished). Thus there is evidence that virus or viruses antigenically related to AHV-1 are prevalent in this subfamily too.

It is therefore probable that there is a family of related gammaherpesviruses which are prevalent in three subfamilies of Bovidae which infect their natural hosts inapparently (Table 1). It would however appear that only two, AHV-1 and the SA-A, transmit to other species of ruminant and induce MCF under natural conditions.

TABLE 1

Distribution of gammaherpesviruses in ruminants

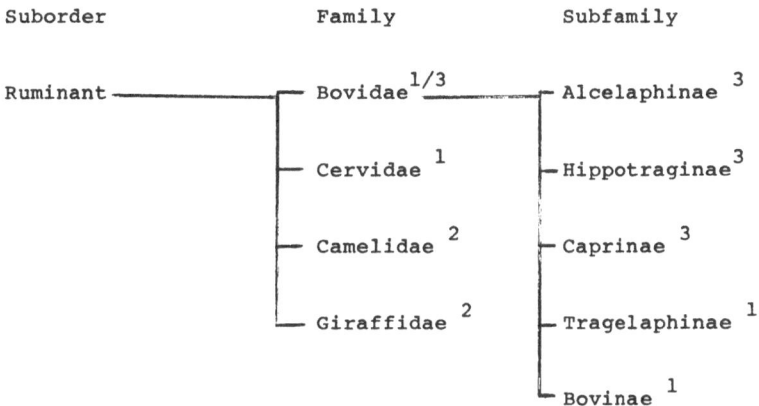

Suborder	Family	Subfamily

Ruminant — Bovidae[1/3] — Alcelaphinae[3]

— Cervidae[1] — Hippotraginae[3]

— Camelidae[2] — Caprinae[3]

— Giraffidae[2] — Tragelaphinae[1]

— Bovinae[1]

[1] Outcome of infection is fatal Malignant Catarrhal Fever

[2] Infection in these families not reported

[3] Evidence of widespread, latent, inapparent infection from serological surveys

CHARACTERISTICS OF THE SHEEP-ASSOCIATED AGENT

Despite concerted attempts worldwide to isolate this virus from sheep and animals reacting with MCF no aetiological agent has been identified. All studies have therefore relied on reagents developed from the prototype strain of AHV-1 (known as WC11) and the detection of cross reactions to it.

Evidence of infection in sheep

Attempts to detect neutralising antibody to WC11 in sheep sera have produced conflicting results (Anon 1972, Rossiter 1981, Harkness, 1985).

However Rossiter (1981) reported that 162/167 sheep sera tested for IIF antibody to AHV-1 were positive and that the only negative sera were 5 derived from gnotobiotic and specific pathogen free lambs. That 9/14 caesarian derived lambs that had been reared in isolation had antibody implies that as with wildebeest a proportion of ovine conceptuses become infected in utero.

In this laboratory we have been able to confirm these observations and extend them by analysing the reaction of sheep serum to AHV-1 by immunoblotting (A.J. Herring, et al., in preparation). These studies indicate that sheep sera react with the 6 major viral components recognised by wildebeest sera although samples from individuals only react with a proportion. This provides compelling evidence that sheep are infected with a virus or viruses that share many antigens with AHV-1.

Evidence from experimentally infected rodents

In contrast to cattle, inoculation of cell suspensions of tissues from deer reacting with SA-MCF, readily transmit the disease to laboratory rabbits (Buxton and Reid, 1980, Reid et al., 1986). Studies of this experimental model have provided much valuable information on the pathogenesis of the disease but no evidence of the nature of the causal virus has been found and like deer reacting with the disease no antibody to AHV-1 can be detected in serum from affected rabbits. It was however found that cell suspensions from rabbits reacting with MCF would transmit the disease to hamsters and disease could then be readily transmitted through hamsters (Jacoby et al., 1988). Reacting hamsters developed a more florid disease than rabbits and with an IIF test it was shown that their sera contained antibody that recognised AHV-1 antigen restricted to the nuclei of infected tissue culture cells (Reid, unpublished). Thus it would appear that the expression of the SA-A in deer and rabbits reacting to infection is insufficient to stimulate an antibody reaction while in hamsters it is sufficent to induce antibody which cross reacts with at least a component of AHV-1.

The importance of this observation is that it provides a link between the antibody to WC11 detected in sheep sera and animals reacting with MCF due to infection with SA-A. The probability that the antibody detected in sheep reflects infection with a virus responsible for SA-MCF can thus be viewed with greater confidence.

Evidence from cells cultured from affected animals

Numerous attempts to isolate virus from animals with SA-MCF have failed to identify the aetiological agent. However from cattle, deer and rabbits reacting with the disease lymphoblastoid cell lines have been cultured that have the characteristics of a specific lymphocyte subset (Reid et al., 1983, 1985).

Cell lines have been propagated from the tissues of both red (Cervus elaphus) and Pere David's (Elaphurus davidianus) deer acutely affected with the disease but cannot be generated from normal deer. In contrast to similar cell lines cultured from cattle with MCF, to date cell lines grown from deer with SA-MCF require a source of exogenous IL-2 for their continued growth in vitro. One such line, MF/816, established on the 30.9.85 from a suspension of lymph node cells from an affected red deer has been maintained in continuous culture by passing every 2-3 days to feeder monolayers of bovine turbinate cells in the presence of 5-10 units of recombinant human IL-2.

All cell lines have the morphology of large lymphocytes and their cytoplasm, in Giemsa stained cytospin preparations, is seen to contain numerous azurophilic granules. In addition these lines express cytotoxicity to a variety of normal cultured cells which is not restricted to histocompatibly identical targets. Thus these lymphoblastoid cells have the morphological characteristics of large granular lymphocytes (LGL) with the functional properties of natural killer (NK) cells. The role of these cells in MCF is still speculative but it is proposed that they are an important target cell for the SA-A and may be crucially involved in the pathogenesis of MCF by causing a profound immunological dysfunction.

This conclusion is supported by observations on lymphoblastoid cells transformed by Herpesvirus ateles and H. saimiri, two simian gammaherpesviruses which produce no disease in their natural host, spider monkeys (Ateles sp) and squirrel monkeys (Saimiri sciureus) respectively (Johnson and Jondal 1981). However when species of Marmoset monkeys become infected with either agent fatal lymphoproliferative disease results and transformed lymphoblastoid cell lines can be cultured from their peripheral blood or affected tissues. The virus-transformed cell lines appear to belong to the same T-lymphocyte subset as the cells generated from SA-MCF affected animals.

There thus would appear to be strong biological similarities between the behaviour of the simian gammaherpesviruses and the SA-A: 1. they produce inapparent infection in their natural hosts, 2. they produce fatal lymphoproliferative disease when infection is established in other species, 3. the pathological changes that develop in animals with lymphoproliferative disease are similar, 4. an important target cell in diseased animals would appear to be the same subset of T-lymphocytes, namely the LGL.

Evidence from molecular studies

Much of the DNA of the WC11 isolate of AHV-1 has now been cloned and an EcoRl fragment representing approximately 15% (20 kilobase pairs) of the genome has been used to probe DNA extracts from the lymphoblastoid cell lines generated from deer, cattle and rabbits with SA-MCF (A. Bridgen et al., unpublished). All cell lines examined appear to contain DNA that hybridises with the WC11 cloned DNA and there is no reaction with normal cells from these species. It is therefore concluded that these lymphoblastoid cell lines carry DNA, homologous to AHV-1, which represents part of the genome of the SA-A.

CONCLUSION

The cause of MCF in farmed deer is the as yet unidentified SA-A. However there is now convincing evidence that it is a virus distinct from, but closely related both antigenically and genomically to AHV-1 of wildebeest. Current efforts to clone the DNA of SA-A from lymphoblastoid cell lines promise to provide specific reagents to examine MCF affected animals and identify virus infected cells. These studies should elucidate the pathogenesis of the disease and assist in the development of improved diagnostic procedures. In addition it is hoped that the use of these reagents to examine sheep tissues will provide evidence about the biology of the virus and identify suitable strategies for the recovery of virus from sheep. Together these achievements will form the basis for the future control of the disease in farmed deer.

REFERENCES

Anon. 1972. Examination of sera from Colorado USA and Greece for MCF
    neutralising antibody. In: Record of Research Annual Report
    1972, E. Afr. Vet. Res. Org. Nairobi, p. 14-15.
Anon. 1980. Animal Health Division of the New Zealand Ministry of
    Agriculture, 7, (2)
Buxton, D. and Reid, H.W. 1980. Transmission of malignant catarrhal
    fever to rabbits. Vet. Rec. 106, 243-245.
Daubney, R. and Hudson, J.R. 1936. Transmission experiments with
    bovine malignant catarrh. J. comp. Path. Therap. 49, 63-89.
Denholm, L.J. and Westbury, H.A. 1982. Malignant catarrhal fever in
    farmed Rusa deer (Cervus timorensis). 1. Clinicopathological
    observations. Aust. vet. J. 58, 81-87.
Gotze, R. 1932. Bosartige Kattarrhalifieber IV. Mitteiburg, Berl.
    Tierartzl. Wschr. 53, 848-855.
Hamblin, C. and Hedger R.S. 1984. Neutralising antibodies to
    wildebeest-derived malignant catarrhal fever virus in African
    wildlife. Comp. Immun. Microbiol. Infect. Dis. 7, 195-199.
Harkness, J. 1985. Bovine malignant catarrhal fever in the United
    Kingdom. State Vet. J. 39, 60-64.
Heuschele, W.P., Fletcher, H.R. Oosterhuis, J., Janssen, D. and
    Robinson, P.T. 1984. Epidemiological aspects of malignant
    catarrhal fever in the USA. Proc. US Anim. Hlth. Ass. 88,
    640-651.
Jacoby, R.O., Reid, H.W., Buxton, D. and Pow, I. (1988). Rodent
    models of malignant catarrhal fever. 1. Transmission of
    wildebeest-associated and sheep-associated disease. J. comp.
    Path. (in press).
Johnson, D.R. and Jondal, M. 1981. Herpesvirus-transformed cytotoxic
    T-cell lines. Nature, Lond. 291, 81-83.
McCallum, H.J.F., Mavor, N.M. and Hemmingsen, P. 1982. A malignant
    catarrhal fever-like disease in red deer (Cervus elaphus) in
    New Zealand. N.Z. vet. J. 27, 230-232.
Mettam, R.M.W. 1924. Snotsiekte in cattle. 9th ed. 10th Resp. Dir.
    vet. Educ. Res. 1923. Union of South Africa, 395-432.
Mushi, E.Z., Rossiter, P.B., Jessett, D. and Korstad, L. 1981.
    Isolation and characterisation of a herpesvirus from topi
    (Damahiscus korrigum, Ogilby). J. comp. Path. 91, 63-68.
Odend'Hal, S. 1983. Malignant catarrhal fever. In. The Geographical
    distribution of animal Viral Diseases. Lond., N.Y. Academic
    Press. p. 277-280.
Plowright, W. 1965. Malignant catarrhal fever in East Africa.
    1. Behaviour of the virus in free-living populations of blue
    wildebeest (Grogon taurinus taurinus, Burchell). Res. vet.
    Sci. 6, 56-68.
Plowright, W. 1967. Malignant catarrhal fever in East Africa.
    III. Neutralising antibody in free livig wildebeest. Res. vet.
    Sci. 8, 129-136.
Plowright, W. 1986. Malignant catarrhal fever. Rev. sci. tech. off.
    Int. Epiz. 5, 897-918.
Plowright, W., Ferris, R.D. and Scott, G.R. 1960. Blue wildebeest
    and the aetiological agent of bovine malignant catarrhal fever.

Nature, Lond. <u>188</u>, 1167-1169.

Reid, H.W., Buxton, D., Corrigall, W., Hunter, A.R., McMartin, D.A. and Rushton, R. 1979. An outbreak of malignant catarrhal fever in red deer (<u>Cervus elaphus</u>). Vet. Rec. <u>104</u>, 120-123.

Reid, H.W., Buxton, D., Pow, I., Finlayson, J. and Berrie, E.L. 1983. A cytotoxic T-lymphocyte line propagated from a rabbit infected with sheep-associated malignant catarrhal fever. Res. vet. Sci. <u>34</u>, 109-113.

Reid, H.W., Buxton, D., Berrie, E., Pow, I. and Finalyson, J. 1985. Culture of malignant catarrhal fever agent. Roy. Soc. N.Z. Bull. <u>22</u>, 143-146.

Reid, H.W., Buxton, D., Pow, I. and Finlayson, J. 1986. Malignant catarrhal fever: Experimental transmission of the "Sheep-associated" form of the disease from cattle and deer to cattle, deer, rabbits and hamsters. Res. vet. Sci. <u>41</u>, 76-81.

Reid, H.W., Plowright, W., Rowe, L. 1975. Neutralising antibody to herpesviruses derived from wildebeest and hartebeest in wild animals in East Asia. Res. vet. Sci. <u>18</u>, 169-273.

Rossiter, P.B. 1981. Antibodies to malignant catarrhal fever virus in sheep sera. J. comp. Path. <u>91</u>, 303-311.

Conclusions from Session IV

More emphasis should be given on the specific infections of deer. It seems obvious now that red deer and other deer are infected by their own specific herpesvirus which simply share some common antigens with the bovine herpesvirus, infectious bovine rhinotracheitis (IBR). Furthermore it was clear that the red deer herpesvirus could not infect cattle while the reindeeer virus did not spread to cattle. The same is true for other diseases of bacterial or parasitic origin e.g. dictyocaulus. Therefore we should recommend at this point urgently to improve diagnostic procedures in order to be able to discriminate between the different infections (e.g. infection with the red deer herpesvirus and IBR). This conclusion is particularly obvious when we want to improve the trade of animals within the European Community. Whenever we want to legislate for animal movement because of veterinary measures we must be sure of our diagnosis. Moreover it is important to reduce the negative impact given in deer farming because of health concern for cattle. As far as other diseases are concerned (for instance MCF) more research is necessary since they can cause major losses in deer farming. Last but not least a big effort should be made to give advice to potential deer farmers or veterinarians in the european countries other than UK deer farming is just beginning, just to avoid real mistakes and avoid duplication of research effort.

<u>SESSION V</u>

Chairman: J.A. Milne

THE TRANSPORT OF DEER

T J Fletcher

Reediehill Deer Farm
Auchtermuchty
Cupar
Fife KY14 7HS
United Kingdom

ABSTRACT

The welfare of livestock, in Britain at least, has become increasingly topical, particularly within the veterinary profession whose obligation to assist its clients in increasing the efficiency of production can run counter to its obligations to safeguard the welfare of the animals.

Transport is considered one of the most stressful procedures in the farm animal's life and has recently therefore been the subject of much debate. Deer are thought to be more susceptible to stress than other farm animals and their image in the public eye is such that any manipulation is controversial. The transport of deer is therefore of relevance to this conference.

This paper outlines the history of deer transport and discusses the current need for transport of deer by road and air. The commonest reasons for casualties during and after movement are considered and recommendations to reduce these are given. Minimum floor areas and ceiling heights are suggested and a list of diseases to which transport stresses might pre-dispose is given.

Today's sessions are devoted to the 'management' of farmed deer and there is no doubt that management faults, certainly in British deer farming, are responsible for far more deaths and for far more loss of production than are any specific disease entities.

Good livestock management has been defined as simply "the minimalising of stress" and, as transport is often considered one of the most stressful of procedures in the life of farmed animals, it is clearly a subject of importance.

It is the view of most deer farmers that deer do travel extremely well; probably much better in fact than other farm animals. Nevertheless they pose special problems different from those of other livestock and this paper will emphasise these and explain how stress can be minimised in the transport of deer.

As many speakers at this meeting have already mentioned, "stress", is crucial in the origin of most, possibly even of all, diseases. Those most significantly presenting as post transport stress in Scotland are: Yersinia pseudo tuberculosis; lungworm; red deer herpes virus; post capture myopathy; winter death syndrome (inanition syndrome, frank starvation); traumatic injury; malignant catarrhal fever.

Yersinia pseudotuberculosis is encountered in apparently healthy animals (Fiona Stuart, this conference) but emerges as a clinical entity particularly in newly weaned calves or recently captured deer. In both situations transport usually precedes the event together with a change of diet, and as the deer are normally housed at this time there is scope for rapid cross infection and losses can be substantial. Administration of Terramycin Long Acting intramuscularly is very effective, provided treatment is given early to all in-contact deer.

Lungworm remains without doubt the major cause of lost production on British deer farms and where deer carrying heavy infestations with few clinical signs are subjected to transport and the stresses of a new environment such infections can prove fatal.

Where the red deer herpes virus has caused clinical disease it has always been associated with transport. Whether the actual stresses of handling are responsible or whether handling before and after movement together with the change in environment is responsible is not known.

Post capture myopathy has been described among African ungulates as well as as in deer which have been captured, usually after pursuit. Transport can also be involved in this condition and among 1,000 wild deer captured by tranquilliser rifle in Britain perhaps one or 2 can be expected to demonstrate the classic wry neck posture associated with post capture myopathy. The prognosis is poor.

It is clear that the syndrome variously described as winter death syndrome, inanition syndrome or frank starvation, is created by the depletion of energy reserves. Consequently transport with its concomitant period of reduced food intake and increased energy output will force animals that were previously marginal into the syndrome. This is certainly true of newly caught deer, often emaciated, which may succumb after transport.

Traumatic injuries do still occur during transport including loading and unloading. Frequently the victim will only be identified on unloading with a fractured femur for example. Occasionally when over crowded the animal will present in a state of shock with crush injuries identifiable at post mortem as multiple fractured ribs due to trampling. This does not occur in the absence of over-crowding.

The incidence of malignant catarrhal fever is known to follow a seasonal pattern in New Zealand deer farms which follows the incidence of nutritional and climatic stress. Very often transport can be the exciting influence in such cases.

Since this is a European gathering we should perhaps quickly mention the history of international travel for deer. The fallow deer was eliminated from most of Europe during the last glaciation and subsequently re-introduced by the agency of man. Thus on the isle of Rhodes, myths indicate its reintroduction in the time of classical Greece to "control the serpents". Probably the Normans brought fallow deer to Britain. By 1230 they had reached Yorkshire in Northern England, by 1244 County Wicklow in Ireland and by 1290 Scotland. In 1608 King James of England imported fallow deer from the King of Denmark. And although Russia is not yet a member of the E.E.C. deer were traded, thus in 1705 Sir Nicholas Le Strange recorded on 11th December that a ship had landed after being run aground in a storm. She carried 50 crated red deer "a gift from the King of Russia to the Duke of Newcastle". More recently and before we had come to fear their imports of lamb and butter, Britain used to export live deer to New Zealand. Many went out by sailing ship to be carried to their eventual release point by wagons drawn by horses, traction engines and even oxen. The red deer prospered and now one in 10 farmers in New Zealand farms deer so that we suffer the importation of their farmed venison at a price below the cost at which we can produce it.

Within Britain and no doubt elsewhere in Europe the movement of deer was commonplace in the mediaeval period. Interestingly there are several accounts of deer being led or herded from one park to another. For example, Joseph Watson, the deer keeper at Lyme Park in Cheshire in the late seventeenth and early eighteenth centuries, drove 24 red stags from Lyme Park to Windsor, about 180 miles, and so won himself a £500 wager. More routine was the driving in 1830 of the entire herd of fallow deer from Watlington Park the 5 miles to Stonor Park, or that a little later of fallow from Syston Park to Revesby Abbey in Lincolnshire.

Nowadays, however, the most stressful of deer movements in routine deer farming is certainly that of the newly caught deer. Scotland's 300,000 or so wild red deer represent the largest source of red deer available for stocking new deer farms in Britain and Europe, and during the last few years increasing numbers of hinds and followers have been taken off the hills. At the moment perhaps 2,000 are being removed annually but the figure is still increasing as the value of a live hind is at least twice that of a dead one and is likely to remain so for many years in the light of demand for breeding stock for new deer farms. It is almost certainly the loading and transport of newly caught wild deer which poses the greatest risk to animal welfare.

It is necessary first to understand the context from which these animals are drawn. In the late winter, that is the time at which deer can be caught by feeding into fenced enclosures, all classes of Scottish wild red deer are likely to be in nutritional stress. The number which actually succumb in a given year depends on the weather during the previous summer as well as conditions during the winter. In parts of the Highlands 25% of the total deer and 75% of the calves may die in a hard winter (Mitchell, McCowan & Parish, 1971).

It follows from the poor condition of the population in general that deer taken from the wild require extremely careful handling. First and foremost the period of extreme stress must be kept as short as possible. This requires that, once enclosed, the animals should be moved quietly and yet as quickly as possible into a vehicle and thence into a building with abundant good tempting food and water. In practice the original enclosure is often only accessible by tractor or four wheel drive vehicle and the deer will require to be loaded into a trailer, taken to a collecting point, held for as short a time as possible before loading onto/ and dispatching to a farm. This last road journey also requires to be as short as possible and should be to a farm where suitable buildings and feed are available in which to house the deer until spring when they can be turned out onto grass. Since the eventual destination is often the south of England, a staging post in Scotland in which the animals can be treated to remove warbles and other parasites and over-wintered or at least held for a few weeks is desirable. Even after this complicated series of manoeuvres mortality of caught deer is usually substantially less than had they been left on the deer forest, and once established on the farm mortality is unlikely to be greater than 2% per annum.

Until about 1970 it was assumed that deer benefitted from individual crating in the time honoured way. As late as 1971 a speaker from the Royal Veterinary College was advocating the use of individual crates with heavy sedation for the regular movement of deer in an address to the

British Deer Society. Of course this was a tradition based on the
necessity of creating a load that could be readily moved by manpower and
horsepower. The advent of fork lift trucks has meant that even for air
and sea transport, large crates holding a number of deer are feasible and
from the point of view of, sociable animals like red and fallow deer, also
desirable. For the more routine transport by road the farmed species,
red and fallow, have been found to be most adaptable and there is no doubt
that deer travel extraordinarily well.

Of the several thousand deer exported I have never yet had any
casulties during transit. Mystique about the movement of deer is misplaced
and common sense plus experience gained with other species of farmed
animals forms a sound basis for transporting deer.

The factors most likely to cause problems are, in order of priority:
overcrowding; mixing different sizes and categories of deer; unsuitable
bedding; overexcitement during loading; high speed cornering and braking;
dehydration; unsuitable vehicles; excessive length of journey.

Dealing with overcrowding first. Specific recommendations of square
footage per head for different categories of deer are not intended as
legally enforceable but it is hoped that those planning to move deer,
especially on long hauls where freight costs can tempt people towards overcrowding will take into account these suggestions (Table 1). Incorporated
into the Welfare Codes, such recommendations may guide those considering
prosecuting farmers and hauliers on welfare grounds.

TABLE 1   Suggested floor areas per head for transport of deer

Category	Floor areas, m^2
Adult stags (antlers removed)	0.70
Adult hinds	0.40-0.60
Yearling stags	0.40-0.60
3 month old calves to yearling hinds	0.30-0.50

I have not attempted to specify maximum floor areas but there are
occasionally situations in which too much space has contributed to
casulties and it is always worthwhile restricting deer to perhaps 3 times
the recommended minimum area so that animals cannot be thrown about and

when loading and unloading cannot damage themselves by running too fast.

So much for floor areas, what about height? Double decked lorries can be used very satisfactorily. Obvious considerations are that during the whole process of loading, travel and unloading there must be no opportunities for animals to jump out. The great agility of deer means that they can go up and down ramps to top decks with great ease and rapidity, and as some farmers have found to their cost, they can equally easily jump the side gates of the normal livestock ramp and trot off. It is not easy to generalise about heights. Increasingly the advent of the long legged east European and north American subspecies of red deer means that heights perfectly adequate for British red deer maybe inappropriate for other strains. Fallow does can manage satisfactorily at as little as 1.1 m and indeed I believe they gain security and are more content in such spaces. British hinds will fit comfortably under a 1.2 m ceiling but even yearling stags can require up to 1.8 m. These figures can be only a guide. Suffice it to say that there must be ample room for deer to stand with head and ears erect. Where the ceiling is low there should be a corresponding increase in allocation of floor space to permit deer to lie down. On a long journey both red and fallow will, of course, lie down but in my experience fallow frequently lie much sooner and therefore seem to travel more contentedly. They are therefore happier under a low ceiling and I believe that this is preferable to one that is too high.

Dividing deer into different livestock classes is essential for transport. Firstly hinds: there is no need to divide large from small adult hinds unless there is clear evidence of belligerence. However, calves of less than 8 or 10 months are probably better separated from their hinds unless there is a very substantial increase in floor space per head. In general, separation of calves from yearlings and yearlings from adults is highly desirable as there is the possibility of the smaller deer being intimidated by the larger and eventually lying down and being trodden on. For the same reason any animal which appears to be the whipping horse for the rest of the group is often better withdrawn.

Where stags are being moved it is vital to remove hard antelrs, or if the stags are in velvet to pen them individually. In Britain no one is permitted to remove velvet antlers, even veterinary surgeons using anaesthetic, and this is true even where it may appear to be in the best interests of the animals and when transport is to take place. In such

circumstances i.e. in velvet stags must then be penned separately with
plenty of space to turn around or where this is not possible they must be
given abundant floor space, say 33 m^2 per head and be transported in very
small groups. When not in velvet stags, even yearlings, must have their
antlers removed and can then be readily transported in groups except for
adult stags around the rut. It will generally be obvious to the farmer
when stags are still fighting during the rut and if there is no sign of
aggression in the pens prior to loading it is unlikely to manifest itself
on the lorry or in the crate, indeed it will almost always be impossible
to persuade the subordinate stag to enter the lorry when there is a
dominant stag within it at that time of the year. Where for example a
group of 2 year old stags are being moved in the autumn it would be a
worthwhile precaution to avoid giving them too much space in which to
contemplate a set to.

Bedding requirements for deer are similar to those of other farm
animals. Straw, woodshavings, peat or sawdust are all adequate. The
purpose of bedding is of course to prevent the crate or vehicle floor from
becoming slippery, to absorb urine and faeces, and to make it more
attractive for deer to lie on. In New Zealand and elsewhere artificial
rubber mats are frequently used instead of bedding and provided they do
not become slippery they are perfectly adequate for at least short journeys.
In the case of sawdust care must be taken to avoid it being more than say
5 cm deep since animals lying down may inhale the dust and can even
suffocate, but a good covering of perhaps 1 cm is nevertheless essential
to provide grip for animals standing in a swaying lorry. On a long
journey hay may be considered a good bedding material as the deer will
eat it and for that reason a plentiful supply is then needed. Different
countries may have a requirement for different types of bedding as part
of their disease prevention protocol and this should be borne in mind
when deer are being exported.

Overexcitement at loading must be borne in mind as it can contribute
very seriously to dehydration and may mean that the animals never really
relax on the journey. It is therefore well worthwhile gathering deer into
pens a day or so before actually loading them. This will also enable
particularly troublesome individuals to be removed where feasible. In
any case deer should be properly fed and watered before loading and any
sorting out into groups or veterinary procedures should be carried out

the day before despatch.

Good loading facilities are obviously crucial and should be designed so that full sized cattle floats can load as well as smaller vehicles. For this reason some farmers have fitted their loading bays with doors having 2 sets of hinges so that the gates can be fitted snugly against the side of any width of vehicle. Doors which can be used to push deer right up against the vehicle ramp are also a great asset, especially if as does occasionally happen, an individual animal is refusing to enter the lorry. A problem seen with stags particularly at rutting time. The only solution here is to take several deer off, mix them with the nervous one and try again. The use of sheets of plywood to walk the offender on is a great help. In all cases it cannot be over-emphasised that quick and quiet loading, as with all deer handling, is vital.

In the case of loading fallow deer the problem is not persuading the deer to board the lorry so much as stopping them jumping off again. There often seems to be only 2 seconds in which to close up the vehicle before the deer are vaulting back out. The solution here is to have a door ready with a fast operator concealed behind waiting to close it. Once in the lorry fallow travel more quietly than red deer.

High speed cornering and braking can be disasterous for all animals; probably deer can put up with this better than cattle. Nevertheless, it is completely inexcusable and brands the driver as incompetent to handle livestock. High speeds are perfectly in order on a motorway but the good driver will drive very slowly on other roads particularly in traffic where sudden stops are possible and unpredictable. The suggestion has been mooted that drivers be licensed to carry livestock, this might be beneficial but in the final analysis it is up to the farmer to assess the driver's skills. As is increasingly the case in suburban fringe farming areas the driver may have little or no experience of livestock and in such situations training courses for livestock hauliers may be a good idea.

Dehydration is the dangerous consequence of all the previous hazards. The metabolic rate of deer is always high and where there is overcrowding, agression between animals, excitement exacerbated by slippery floors, bad loading techniques, fast cornering or sudden braking, then panting may occur, especially in high temperatures. Poor ventilation can rapidly create a high temperature and it is always well worthwhile moving off as soon as possible after loading and then stopping later to check that all is well and perhaps to reduce air flow by closing vents. Panting of course leads to rapid loss of water and deer, even though they are ruminants with the capacity of up to perhaps 20 litres of moist food in their rumen alone, can soon suffer dehydration. For this reason in hot weather spraying with water whether by provision of a sprinkler system in the lorry or simply by using a hose through the vents is often appreciated by deer. If the vehicle is stationary for long then watering can be done by bucket which will require constant replenishing and will need to be placed at strategic intervals throughout the vehicle. Where deer are accustomed to roots or other moist foods these are ideal for maintaining water balance but a plentiful supply will be needed. Care must also be taken when exporting animals that the very presence of the feed does not constitute an infringement of some countries' health status. Thus I well remember how the thoughtful provision of apples to deer travelling to New Zealand created an alarming international incident.

Where aircraft are forced to stand on airfields it is imperative that the airport authorities are prepared to connect air conditioning units as soon as the plane lands. These essential air conditioning systems exacerbate dehydration and huge volumes of water are required for long haul flights, of the order of 10 litres per head. This can perhaps best be supplied by means of buckets fitted into plywood panels in the corners of crates so as to avoid trampling and spillage. They will then need to be filled by the flight attendant using a stirrup pump.

Probably the most important means of limiting dehydration is to provide moist food in plenty in the hours preceding departure. Given a bellyful of roots or green food deer will often resist drinking for 12 hours although attempts to water by hose or bucket should be made before this.

On arrival, where deer appear to be dehydrated care should be taken to prevent sudden rehydration, and even more importantly, great care is necessary in allowing access to 'concentrates' such as corn or pelleted rations. Deer are very susceptible to rumenal acidocis and several deer have been killed by kindness after a long journey. The ones at risk are usually the tamest, those which are accustomed to concentrates and of course those which have not been fed on the road. Deer should always be fed forage and perhaps roots or other succulent food after arrival until they can gradually be reintroduced to concentrates.

Deer must have been transported in a greater variety of vehicles than almost any other animal. In general those suitable for cattle and sheep will prove ideal for deer provided that the partitions are solid enough to prevent animals squeezing through and high enough to stop them jumping over. As mentioned before, a reasonably low ceiling is helpful. There is no advantage in having a darkened vehicle and indeed plenty of ventilation through open side vents is often helpful although in cold weather this should be reduced especially if the vents are too near eye height. Ideally vents at knee height and above the head are ideal. A current of air blowing into the eyes of deer can produce a transient conjunctivitis.

And now the question of journey length. Properly loaded and driven deer, with suitable floor area and ventilation, adequately fed and watered are best left on the lorry rather than being subjected to the antiquated concept of 'lairage'. Loading deer on and off for a few hours in an alien environment seems to me entirely counterproductive. In all cases I believe it is greatly in the animals interest to be moved as quickly as possible on these journeys with stops kept to an absolute **minimum**. In practice the delays are always those at the border crossing. So often while waiting for a border veterinary inspector to sign papers whilst scarcely glancing at the animals it has seemed dreadfully ironic that it should be the veterinary procedures which prolong the deer's journey.

EARLY NUTRITION, GROWTH AND REPRODUCTIVE PERFORMANCE IN YOUNG SCOTTISH RED DEER (CERVUS ELAPHUS) HINDS, AND THEIR ECONOMIC SIGNIFICANCE IN COMMERCIAL HERDS

W J Hamilton

Macaulay Land Use Research Institute
Glensaugh Research Station
Laurencekirk, AB3 1HB

ABSTRACT

The importance of an adequate level of nutrition to permit high growth rates in calves and young hinds up to 16 months of age, in order to achieve high levels of lifetime reproductive performance is highlighted. The positive relationship between the liveweight of the hind at the rut and a number of parameters of subsequent production is also demonstrated. The effects on deer production on a west Scotland farm of the implementation of an improved system of management for both livestock and the pastures, are described. In three years mature hind liveweights increased by 11.8% to 87 kg, weaning weights of replacement hind calves increased by 32% to 42 kg, with the summer growth rates of the young hinds on sown pastures increasing by 51%. The reproductive performance of the s.e. young hinds increased from a mean over the previous six years of 7% to 83%. The failure of a young hind to calve at two years of age was estimated to reduce the gross margin of £45 per hind by 16%. The recruitment of successive cohorts with similar levels of performance reduced the annual gross margin by at least 22%. The economic effect of similar levels of production on a single cohort lowland enterprise was shown to be considerably more serious.

INTRODUCTION

The proportion of young red deer hinds which calve at two-year of age on U.K. deer farms varies from 0 to 1.0. Scottish farmed red deer will perform at a high level for up to 14 calf crops and probably beyond (Hamilton, Milne and Maxwell, 1984) with a lifetime performance over that period of 0.98 successful pregnancies and a weaning rate of 0.90. Well-established herds are usually comprised of regularly aged cohorts of hinds recruited each year to replace culled hinds. On such farms or those farms where hinds of several ages have been purchased, the economic effect of a poor level of performance by the young hinds is reduced, because of the small proportion of young hinds entering the herd per annum (0.08) (Hamilton W J, 1986). In recently established farms the breeding herd could consist of a single cohort purchased as young hinds at the start of the enterprise. In these circumstances the performance of red deer, which are not well grown and which fail to breed and reproduce in their adult year (16 to 24 months), would have a serious effect on the economic viability of the enterprise in that first year and on the return on the capital invested in the longer term. This paper will consider the management of red deer calves and young hinds in order to ensure a high level of performance at an economic cost.

Relationships between nutrition in early life and reproduction

The relationship between nutrition and growth in early life and subsequent reproductive performance has been described by Hamilton and Blaxter (1980), Blaxter and Hamilton (1980) and Blaxter, Boyne and Hamilton (1980). The fertility of the hinds was related to their bodyweight at the time of the rut. Smaller hinds calved later, and within the heavier hinds in the herd, calving was earlier by one day for each additional 4 kg of liveweight. The data also demonstrated that hinds weighing 52 kg at the rut or less will not calve at all and the calving probabilities of hinds weighing 60 and 80 kg were 0.49, and 0.91 respectively. Hinds, weighing 87 kg or more, all produced calves. These relationships were obtained from an analysis of the performance data of cohorts of hinds, which had been fed at different levels of nutrition during their first winter. The calves were all outwintered on a heather-dominant hill pasture and were fed different amounts of a concentrate supplement (ME = 11 MJ/kg) in each of 5 years and offered hay ad-libitum in snow-storm conditions. The concentrate supplement was fed from weaning at the end of September to the end of May, when the calves were 5 to 11 months of age. Subsequently the hinds remained on hill pastures for several years. They were fed hay only in winter storm conditions and were fed a total of 28 kg per head of the same concentrate in late pregnancy and early lactation during May and June each year. The liveweights, feed inputs, growth rates and subsequent reproductive performance of the cohorts are detailed in Table 1.

TABLE 1 Liveweights, concentrate feed inputs, and reproductive performance of young hinds

Cohort /year birth	Liveweight at weaning (3 months) kg	Concentrate fed per head per day in winter kg	Liveweight in April (9 months) kg	Liveweight in September (15 months) kg	Calving Rate (24 months)
A-1970	34.2	1.0	49.3	74.4	1.00
B-1971	30.0	0.91	46.4	66.5	0.98
C-1972	32.6	0.68	42.5	64.9	0.81
F-1973	30.0	0.57	35.4	51.1	0
H-1974	32.2	0.45	40.3	48.8	0

The growth rate of the 'F' and 'H' cohorts was so severely affected by their early nutrition that only 70% of the F's and 43% of the 'H' cohort calved at 36 months. The consequences of low levels of nutrition and poor growth in early life extended beyond the low reproductive perfomance of the young hinds; it was also reflected in the birth weights of their calves, and growth rates of calves to weaning. Birth weight was found to be significantly related to the weight of the hind at the rut, as was the weight of the calf at weaning and its weight at 16 months. This means that a calf born to a small hind is disadvantaged in three ways. It will be smaller at birth, later born, and with a lesser growth rate and with a shorter season in which to grow will be smaller in September at weaning than a calf born to a heavier hind. A hind weighing 85 kg at the time of the rut can be predicted to produce a hind calf weighing 7.5 kg at birth, and which will weigh 40 kg at weaning in September and, given an adequate level of nutrition, will weigh 75 kg at 15 months. Young hinds of this size will have a high probability of breeding at 16 months of age.

The growth and reproductive performance of young hinds on a commercial farm

The importance of early nutrition and growth on the subsequent breeding success of young red deer hinds in a commercial herd is illustrated from the performance data of young hinds on the Highlands and Islands Development Board's Rahoy deer farm between 1977 to 1982.

These young hinds were reared on a similar plane of nutrition over winter to that of the 'A' cohort in Table 1 but were grazed in the summer on predominantly sown grass species. The Rahoy deer farm, which is situated in the Morvern peninsula in Argyll in west Scotland, extends to some 686 hectares of rough hill grazings, 32 hectares of improved hill grazings and 20 hectares of grassland. The land resources rise from sea level to 462 m. The farm enjoys a mild maritime climate with a rainfall of approximately 2000 mm per annum. The farm currently carries a deer stock of 650 breeding hinds and 40 mature stags. A total of 60 young hinds are recruited to the herd each year. The farm was established by the Highlands and Islands Development Board in 1977.

The very poor levels in performance of the young hinds, can be seen from the data presented in Table 2. The liveweights of mature hinds (78 kg) at Rahoy over the same period were considerably lower than those observed in hinds associated with the study described previously (88 kg). The low level of reproductive performance of the young hinds was related to their liveweight at the time of the rut, which was a consequence of their low weaning weights and growth rates to 15 months of age.

The overall performance of the deer at Rahoy up to 1982 was therefore well below the potential for the species as demonstrated in the first study. In 1983, in order to allow the deer stocks to perform to their full potential, a system of management was devised for the farm for the management of the deer and the pastures on a year-round basis. The principal features of the system were as follows:-

1. Calves were weaned at the same time (21 September) each year.

2. Female replacement calves were housed at weaning and fed a 16% protein concentrate and barn dried hay ad-libitum for a period to two weeks. Weather permitting, they were returned to swards with a surface height of 8 to 10 cm.

TABLE 2 The liveweight and reproductive performance of young hinds at Rahoy from 1977 to 1982

Year of Birth	Liveweight at weaning (3 months) kg	Liveweight in April (9 months) kg	Liveweight in September (15 months) kg	Calving Rate at 24 months
1977	-	39	61	0.14
1978	29	39.5	58	0.06
1979	28	45	58	0.08
1980	32	45	57	0.01
1981	32	46	53	0
1982	31	44	58	0.16

The calves remained at pasture with a daily concentrate supplement of 0.25 kg/hd until the grass was fully utilised or bad weather intervened, usually during November, when the calves were re-housed. Feed levels thereafter were as follows:-

(a) Up to 21 December - 1 kg concentrate + hay ad libitum

(b) 22 Dec to 28 Feb - 0.5 kg concentrate + hay ad libitum

(c) 1 March to 30 April - 1.0 kg concentrate + hay ad libitum

(d) 1 April until turn-out to pasture - 1.3 kg concentrate + hay ad libitum

Calves were turned out to the pasture when the mean sward height reached 6 cm.

3. The sown grass pastures received regular applications of a compound fertiliser (N.P.K.). With the total amount of Nitrogen applied being 215 kg/ha per season. The sward height in the arable pastures was maintained at

between 6 and 8 cm throughout the grazing period.

4. The young and growing hinds were kept on the best pastures during the summer grazing periods until at least three years of age by which time they were fully grown mature hinds.

5. The improved pastures were cleared of all stock from 1 December to mid-April to prevent poaching and to improve summer grass production.

6. The hinds were wintered in groups on the areas of rough grazing, each with a feeding site.

7. All the hinds received hay and concentrates to minimise any loss of liveweight over the winter period.

8. A preventative medicine programme was devised to keep deer free from parasites and to prevent mineral deficiencies.

The effect of the improved year round nutrition on the growth and performance of the calves from 3 months of age to 15 months can be seen from the data presented in Table 3.

TABLE 3 Liveweights and growth rates of the calves from 3 to 15 months of age for 1983-1986

Year of Birth	Liveweight at weaning (3 months) kg	Liveweight in April (9 months) kg	Liveweight in September (15 months) kg	Calving Rate at 24 months
1983	32.1	47.8	62.5	0.43
1984	36.2	44.5	67.1	0.53
1985	40.0	47.8	74.6	0.83
1986	42.4	52.7	75.7	-

Over the period from 1983 to 1986 the weaning weights of the calves have increased by 32% and the growth rates of the young hinds from 103 to 156 g/day over the summer. The birth weights of the calves and the liveweight of the mature hinds have also increased. The birth weights of the calves have increased from 6.6 to 8.4 kg between 1983 and 1986 and mature hinds' liveweights in 1986 were greater by 9.2 kg in 1986 compared to those in 1983. The reproductive performance of the young hinds has also improved dramatically and is now approaching an acceptable level of production.

These results demonstrate the importance of operating a management system which provides adequate levels of nutrition to allow young hinds to reach puberty at 16 months of age and shows the significance of adopting a set of decision rules based on animal production research.

Economic considerations

The gross margin per hind (output (£90) less variable costs (£45)) calculated for Rahoy, when 85 calves per 100 hinds are weaned, is £45 per hind. This figure reflects the high level of performance from the young hinds in the herd. The total failure of the young hinds to breed and to produce viable calves, as happened in 1983, would result in a reduction in the gross margin of £7.20 or 16% to £37.80 per breeding hind. This particular cohort only achieved a calf weaning rate of 63% in 1984 as three year old hinds, which represents a loss of £2.64 per breeding hind or 5.8% of the possible gross margin. The cumulative effect of recruiting small young hinds into a herd in successive years would have the effect of reducing the gross margin by at least £9.84 or 21.8%, to £35.16 per hind. At this level, with fixed costs on such a farm of around £30 per breeding hind, the small margin of profit would represent a return on tenant's capital investment of less than 2% per annum. The effect of a similar reproductive failure in young hinds purchased for a single cohort enterprise would be much more serious. The cost of young hinds entering the herd would often be substantially greater than when recruited from home-produced stock, as is normal on hill and upland farms. Apart from the substantial negative gross margin in the first year, the cumulative effect on the lower levels of performance in the third and subsequent years of poorly grown young hinds would be to reduce the potential gross margin for several years until such time as the hinds reached liveweights which would achieve high levels of performance. In enterprises of this kind therefore, it is extremely important to relate the purchase price of young hinds to their liveweight at the rut, and hence the probability of calving at 24 months. The profitability of a deer farming enterprise will depend to a large extent on this price/liveweight relationship and the standard of management applied thereafter to both livestock and pastures.

ACKNOWLEDGEMENT

I am grateful to Mr N.S. Sutherland of HIDB for the use of the data from Rahoy for the years 1977 to 1982.

REFERENCES

Blaxter, K.L., Boyne, A.W. and Hamilton, W.J. 1980. Reproduction in farmed red deer, 3: Hind growth and mortality. J. agric Sci., Camb. 96, 115-128.

Blaxter, K.L. and Hamilton, W.J. 1980. Reproduction in farmed red deer, 2: Calf growth and mortality. J. agric Sci., Camb. 95, 275-284.

Hamilton, W.J. 1986. Deer as an alternative system, Reference. Book and Buyers Guide, Royal Agricultural Society of England 1986.

Hamilton, W.J. and Blaxter, K.L. Reproduction in farmed red deer. 1: Hind and staf fertility. J. agric Sci., Camb. 95, 261-273.

Hamilton, W.J., Milne, J.A. and Maxwell T.J. 1984. Progress in research on red deer farming. HFRO Biennial Report, 1982-83, 159-164.

FARM TOUR OF MACAULAY LAND USE RESEARCH STATION AT GLENSAUGH

A SUMMARY OF THE MAIN FEATURES

J.A. Milne

Macaulay Land Use Research Institute, Bush Estate, Penicuik,
Midlothian, EH26 OPY

Research over the last 15 years has shown that the red deer can be farmed successfully. Work conducted on nutrition, reproduction and grazing management has been synthesised into the development of production systems for the production of weaned calves and of 16-month old animals for slaughter. These systems have been tested successfully at Glensaugh and the Highland and Islands Development Board's Rahoy Farm and are acting as a blue-print for the development of the deer industry in the UK.

However the major disadvantage of such systems is that they produce calves for the market over a relatively short period in the autumn and early winter and in the development of future markets there is a need to spread the supply over a much longer period of the year. Research is therefore being concentrated on

(1)   altering the date of calving from mid-June to April or early May which has the advantage of matching grass production with peak nutritional demand better thus allowing more rapid growth rates of calves over a longer period in the summer and making in possible to slaughter calves at 10-12 months of age. Research has concentrated on the potential for using the Pere David's deer to hybridise with the red deer, since the Pere David's deer naturally calves in April.

(2)   improving growth rates by the use of larger deer species. The Waipiti has been hybridised with the red deer with a 60% success rate and the growth rate of the offspring have been significantly higher than that of the red deer. The Pere David's deer also has potential in this regard since the mature weight of hinds is 60-70 kg greater than that of the red deer.

(3)   improving growth rate of red deer calves in the winter. Most deer have a natural inappetance period in December to February. Current research is examining the mechanisms controlling the seasonal cycle of

appetite with the objective of enabling manipulation of these cycles and thereby improving growth rates.

The advantages of the deer over other domesticated ruminants are the longevity of the adult hind and the ability of yearling females to breed. Research on both aspects is being conducted to improve the biological and hence economic efficiency of deer production systems.

Conclusions from Session V

The transport of deer presented no major problems provided precautions were taken to accommodate the physical, physiological and temperamental characteristics of the species. There was nothing novel in transporting deer as it had been practiced for many centuries. Different species of deer had specific requirements and this had to be observed particularly at points of loading and unloading. Provided the guidelines described were followed deer could be transported over long distances without causing undue stress.

The essential role of adequate nutrition of red deer if they were to meet their reproductive potential was emphasised during this session. Rigorous attention to grass management was emphasised as essential to profitable deer farming. Not only had the hinds to be fed to ensure conception and foetal development but the young stock had to be kept on a plain of nutritioin that ensured that adequate body size was achieved by reproductive age. The profitability of red deer farming under Scottish conditions was superior to that of other livestock enterprises.

During the farm tour numerous developments were described. Of particular value could be the exploitation of the characteristics of Wapiti and Pere David's deer for farming purposes. Investigations to assess the feasibility of developing hybrid stock was an exciting development and should be pursued vigorously.

The meeting was considered unanimously to have been of great value. It provided for the first time in Europe a forum for scientists interested in the development of deer farming to meet and exchange views. Although the industry is in its infancy in Europe many problems associated with the farming of these species have been identified and schemes of management developed to solve them. The dissemination of this information, to ensure that deer farming developed throughout the Community as an economically viable alternative to other forms of livestock husbandry, was identified as a high priority and it was recommended that a further meeting should be held in two years time.

Work on the domestication and development of management strategies for fallow deer had been largely undertaken in Germany while the potential of red deer and related subspecies had been mainly investigated in Scotland. Continued support for these research programmes was clearly essential if the industry was to progress. Mis-management was identified as the principal cause of loss at present but a number of other health problems were considered to be emerging. Strategies for the control of helminth infections were described at the meeting and would appear to be satisfactory although methods for measuring the extent of parasitism had to be further assessed. However certain viral, bacterial and protozoon infections were identified as causing disease which differed from that recognised in other domestic ruminants. These could present problems to individual farmers but in addition compromised the movement of breeding stock between member states. Knowledge was however incomplete with regard to incidence, economic importance, and pathogenesis of these diseases and in addition many presented serious diagnostic problems. Concern was also expressed about the transmission of infection between deer and other domestic ruminants. The UK had played an important role in generating information on infectious diseases of deer and it was clear that there would be an increasing demand for the development of improved diagnostic tests and strategies for control.

The coordination of research efforts into the development of management schemes and health problems was considered essential to avoid wasteful duplication within the Community. The location in Scotland of centres for research into management, health and reproduction as well as the worlds largest source of red deer stock could provide the necessary coordinating focus to promote deer farming throughout the Community.

BELGIUM

P.P. PASTORET
Faculte de Medicine Veterinarire,
45 Rue des Veterinaires
B 1070 Bruxelles,
Belgium

E. THIRY
Faculte de Medecine Veterinaire
45 Rue des Veterinaires
B 1070 Bruxelles
Belgium

DENMARK

R.J. JORGENSEN
Institute of Internal Medicine
Royal Veterinary & Agricultural Univ,
Bulowsvej 13,
DK-1870
Copenhagen V,
DENMARK

H.V. KROGH
National Veterinary Laboratory
Bulowsvej 27,
P.O. Box 373,
DK-1503,
Copenhagen V,
DENMARK

F. VIGH-LARSEN
National Institute of Animal Science
Forsogsanlaeg Foulum
Postboks 39,
DK-8833, Orum Sdrl,
DENMARK

FINLAND

C. EK-KOMMONEN
National Veterinary Institute,
P.O. Box 368,
01001 Helsinki,
FINLAND.

FRANCE

J. BARRAT,
Le Chef du Service,
Pathologie des Animaux Sauvages
Domaine de Pixerecourt,
B.P. No. 9, 54220 Malzeville,
FRANCE

M. PRAVE
Ecole National Veterinaire de
  Lyon
Marcy l'Etiole, 69620,
Charbonnieres, Les Bains
FRANCE

GERMANY

H. HEMMER
Institute for Zoologie
University Mainz,
WEST GERMANY

G. REINKEN,
Landwintschafts Kammer
Reinland,
Edenicher Allie,
D-53-Bonn
WEST GERMANY

J. WINKELMANN
Robeweg-5211
Tiergesumdheitsamt,
Bonn-3,
WEST GERMANY

IRELAND

M. O'Toole,
A.F.T., Creagh,
Ballinrobe,
Co. Mayo
IRELAND

NETHERLANDS

F.H.M. BORGSTEEDE
Central Vet. Institute,
Edelhertweg 15,
8219 PH Lelystad
THE NETHERLANDS

UNITED KINGDOM

T.L. ALEXANDER
27 Hereford Close,
Minstergate
Off Victoria Road,
Beverley.

D.A. BLEWETT
Moredun Research Institute
408 Gilmerton Road,
Edinburgh, SCOTLAND

M.E. BROWN
Veterinary Investigation Centre,
Quarry Dene,
Westwood Lane,
LEEDS

D. BUXTON
Moredun Research Institute
408 Gilmerton Road,
Edinburgh,
SCOTLAND

J. FLETCHER
Reedie Hill Farm
Auchtermuchty
Fife, SCOTLAND

N.J.L. GILMOUR
Moredun Research Institute
408 Gilmerton Road,
Edinburgh,
SCOTLAND

J. MILNE
MLURI
Bush Estate,
Penicuik,
SCOTLAND

R. MUNRO
Roselea
Broomieknow
Lasswade
SCOTLAND

P.F. NETTLETON
Moredun Research Institute,
408 Gilmerton Road,
Edinburgh,
SCOTLAND

H.W. REID
Moredun Research Institute,
408 Gilmerton Road,
Edinburgh,
SCOTLAND

R.H. SEED
MAFF
Windsor House,
Cornwall Road,
Harrogate

F. STUART
Central Veterinary Lab,
New Haw,
Weybridge,
Surrey.

C.E.C.

J. CONNELL
Commission of the European
   Communities
Rue de la Loi 200,
1049 Brussels,
BELGIUM